# SAT MATH LEVEL 1

## 5 Full Practice Test

for Mathematics Level 1 Subject Test

2017 edition

In Remembrance of

## King Bhumibol Adulyadej

About Author:

Ankur Sharma is an experienced professor who had been teaching math and physics for more than fifteen years at AIMS located in Thailand. He earned a B.Engineering a M.SC in TechnologyManagement from Assumption College with full scholarship. Most of the students who studied SAT Physics and Math had pass the exam easily with a score of 650 above

His goal for writing this book was to help students and examinee to use the time efficiently thru the right material with a proper method.

Updated: June 2017

Feel free to ask any question on facebook page: facebook.com/IGEN4U

ISBN-13: 978-1548997274

ISBN-10: 1548997277

BISAC: Education / Adult & Continuing Education

# Contents :

# Topics on exam

**Numbers and Operations**  (about 10% to 14%)

- **Operation, ratio and proportion**
- **Complex numbers, counting and elementary number theory**
- **Matrices, sequences, series and vectors**

**Algebra and Functions** (about 38% to 42%)

- **Expressions, equations, inequalities, representation and modeling, properties of functions** (Linear, polynomial, rational, exponential, logarithmic, trigonometric, inverse trigonometric, periodic, piecewise, recursive, parametric)

**Geometry and measurement** (about 38% to 42%)

- **Coordinates** (Lines, parabolas, circles, ellipses, hyperbolas, symmetry, transformations, polar coordinates)
- **Three-dimensional** (Solids, surface area and volume , coordinates in three dimensions)

**Data analysis, statistics and probability** (about  8% to 12%)

- **Mean, median, mode, range interquartile range, standard deviation, probability, graphs and plots, least squares regression** (linear, quadratic, exponential)

*credit:*

*https://collegereadiness.collegeboard.org/sat-subject-tests/subjects/mathematics/mathematics-2*

# What's inside SAT Mathematic Level 1 subject test ?

SAT Mathematics Level 1 subject test requires you to have more than three years of college preparatory mathematics, which includes two years of algebra one year of geometry, and precalculus.

You will be tested upon fundamental concepts and knowledge, single-concept problem, and mixed concept problem.

Remember that all the formulas require you to convert everything to SI-base units before using them to obtain a correct result.

# How to calculate the score ?

**Raw score**  =   number of correct answers  -  ( 0.25 x number of  wrong answers)

The raw score is then converted using a curved calculation for overall score into range of 200 to 800.

# Difficulty level of each TEST are stated below using star

Easy :        ☆☆☆

Moderate:    ☆☆☆☆

Advance :    ☆☆☆☆☆

# Tips and Tricks :

**About the test:**

    i.        Calculator is allowed (Not all model are allowed)

    ii.       There are 50 questions multiple choice

    iii.      60 minutes is the time limit

    iv.      1 correct gives 1 point

    v.       1 wrong gets minus 1/4 point

    vi.      A blank gives 0 point

    vii.     Total full score is 800

**Guide-line in taking the test:**

    i.        Time yourself

    ii.       Do easy question first

    iii.      Each question is awarded same mark

    iv.      Pace yourself after 5-6 questions

    v.       Round the numbers for easy calculation

    vi.      Double check your work for units

    vii.     Write the formula out

    viii.    Draw out the diagram for visualization

    ix.      Eliminate choices

    x.       Be confident in yourself

# How to use this book ?

**Guide-line in using this book :**

### 1. *Time is key number one*

This book has a timer that help you pace yourself by looking at the <u>top-right corner</u> you will notice a sand watch.  There is also difficulty level rating in this book.

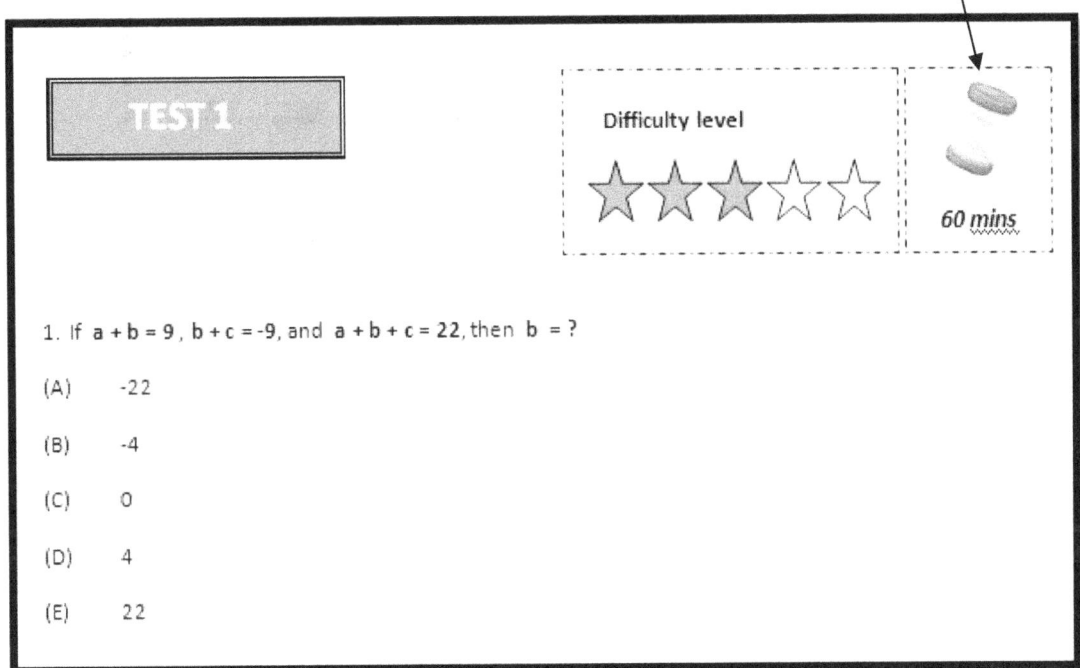

TEST 1

Difficulty level

★★★☆☆

60 mins

1. If $a + b = 9$, $b + c = -9$, and $a + b + c = 22$, then $b = ?$

(A)    -22

(B)    -4

(C)    0

(D)    4

(E)    22

### 2. *Draw the diagram and list the formula out*

The key here is to <u>list what we know</u> to see how to solve the problem then list the formula out

3. What is the sum of the infinite term of the geometric series below ?

$$2^0 + 2^{-1} + 2^{-2} + 2^{-3} + ...$$

(A)    5.5

(B)    2.5

(C)    1.9

(D)    1.5

(E)    0.33

Step 1: List the given info   $U_1 = 2^0 = 1$   and $r = 1/2$

Step2: List out formula for Sum of Geometric Sequence

$$S_{\infty} = \frac{U_1}{1-r}$$

Step 3: Plug in and solve

$$S_{\infty} = \frac{1}{1-1/2}$$

## 3. *Rounding numbers in calculation makes life easy*

Round the numbers for easy calculation using some tricks below

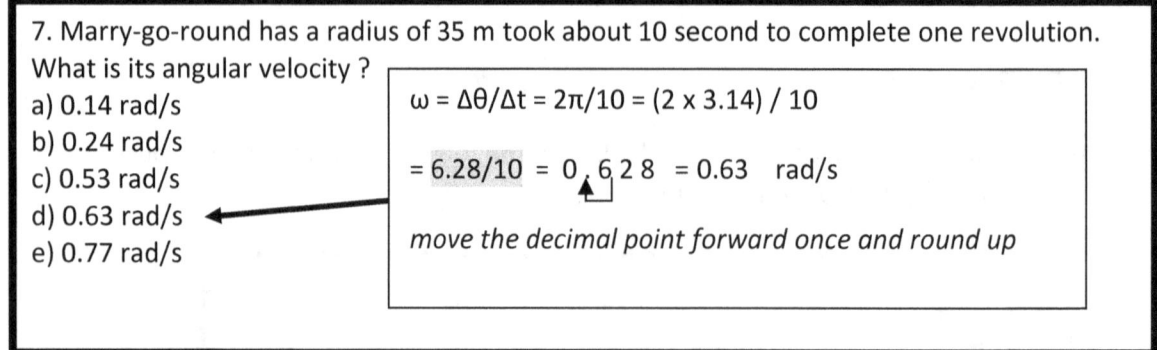

7. Marry-go-round has a radius of 35 m took about 10 second to complete one revolution. What is its angular velocity ?

a) 0.14 rad/s
b) 0.24 rad/s
c) 0.53 rad/s
d) 0.63 rad/s
e) 0.77 rad/s

$\omega = \Delta\theta/\Delta t = 2\pi/10 = (2 \times 3.14) / 10$

$= 6.28/10 = 0{.}628 = 0.63$ rad/s

*move the decimal point forward once and round up*

## 4. *At the end of the test there is a solution key with a <u>detailed explanation</u> and <u>method used</u>, try to understand the question you got wrong.*

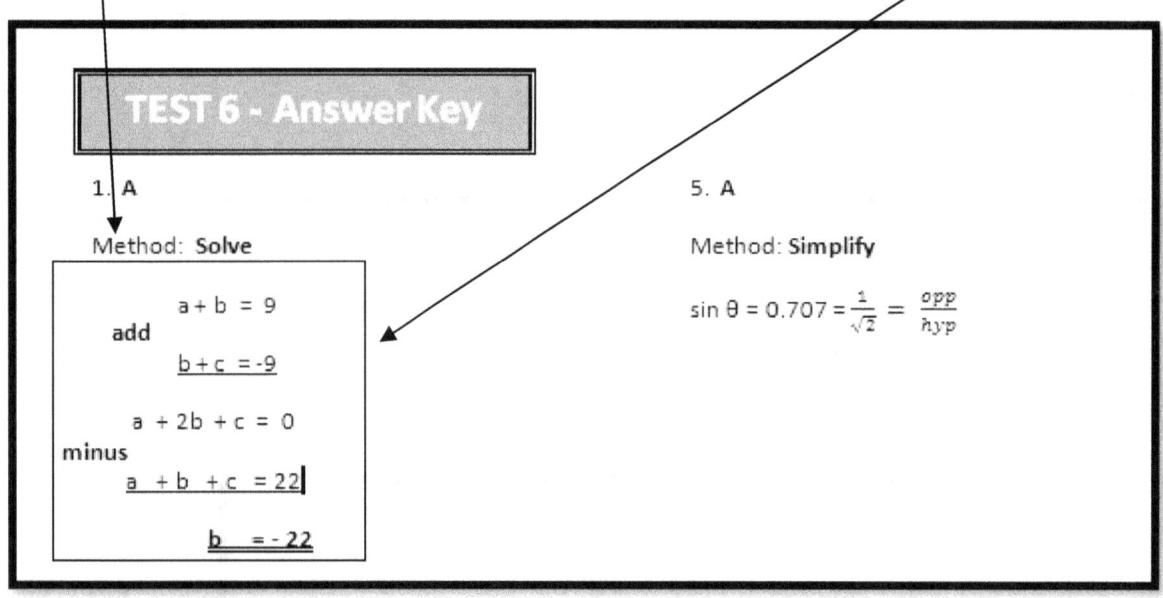

**TEST 6 - Answer Key**

1. A

Method: **Solve**

$a + b = 9$
add
$b + c = -9$

$a + 2b + c = 0$
minus
$a + b + c = 22$

$b = -22$

5. A

Method: **Simplify**

$\sin \theta = 0.707 = \frac{1}{\sqrt{2}} = \frac{opp}{hyp}$

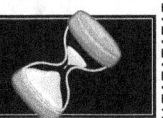

1. If **U = V - 3** , **-V = U + 2** then **U² - V² = ?**

(A)     -6

(B)     -5

(C)     -4

(D)     4

(E)     6

2.   What is the slope of line that passes thru point (-1, 5) and (2, 5) ?

(A)     -0.33

(B)     -1.33

(C)     0

(D)     2.33

(E)     3.33

3.   What is the sum of the first five terms of the geometric series below ?

$$3^0 + 3^{-1} + 3^{-2} + 3^{-3} + ...$$

(A)     40/81

(B)     40/27

(C)     121/27

(D)     121/81

(E)     121/243

4.   What is the value of **a** if **a = $\sqrt[3]{2^2 - 12}$** ?

(A)     $\sqrt{10}$

(B)     10

(C)      8

(D)     -8

(E)     -2

5.   What is the area of a circle with a circumference of 18π ?

(A)     36 π

(B)     64 π

(C)     81 π

(D)     121 π

(E)     324 π

6.   Which of the following is the equation of the line parallel to line y = 2x - 12 ?

(A)     y = 0.5x + 12

(B)     y = -2x + 12

(C)     x = 0.5y +12

(D)     x = 0.2y +12

(E)     x = y

7. If $a(x) = 5^x$ and $b(x) = 5^{-x}$,

then $a(b(0)) = ?$

(A)    $\frac{1}{5}$

(B)    0

(C)    1

(D)    5

(E)    25

8.  If $f(x) = -x^2 + 16$ and line L is tangent to the graph at its vertex then what are the x-intercept of the graph ?

(A)    -16    and    16

(B)    - 12    and    12

(C)    - 8    and    8

(D)    - 4    and    4

(E)    -2    and    2

9.  Which of the following graph of a function ?

 (A)    x = 1

(B)    y = 1

(C)    x = 0

(D)    $y = \pm \sqrt{x}$

(E)    $x \cdot y^{-4} = 1$

10. The isosceles triangle below has a height of 12 meter

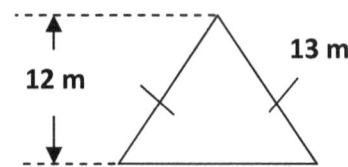

What is the perimeter of the triangle?

(A)    33

(B)    35

(C)    42

(D)    46

(E)    48

11.  If $\log_a 9 = b$ , $\log_a 2 = d$  and

 $\log_a X = b + d$  then  X =

(A)    72

(B)    36

(C)    18

(D)    9

(E)    4

SAT MATH LEVEL 1 Practice-Test

12. If the radius of a circle is doubled then by what percent would the area increased by ?

(A)     40 %

(B)     100%

(C)     200%

(D)     300 %

(E)     400%

13. When IBM produces a computer the cost of production is $250 on each unit.

It sells each unit for $ 480. What is the profit made by IBM for selling 20 units of computer ?

(A)     3600

(B)     4600

(C)     5000

(D)     7000

(E)     9600

14. If $2x - 3 = ( x - 1)^2$ , then $x =$

(A)     -1  only

(B)     0 and -1

(C)     2 only

(D)     2 and -1

(E)     0 and 2

15. How many ways can 5 students arrange themselves in a straight line if the class president must be at in front ?

(A)     9

(B)     16

(C)     24

(D)     96

(E)     120

16. What is the period of the graph
$y = 2 \sin(2x) - 3$ ?

(A)     $\frac{\pi}{6}$

(B)     $\frac{1}{6}$

(C)     6

(D)     $\pi$

(E)     $6\pi$

17. Given that $y = \dfrac{e^x}{x-2}$ , what is the value of $x$ that will make $y$ undefined?

(A)     0

(B)     2

(C)     $e$

(D)     $e^2$

(E)     $\infty$

18. If $g(x) = -x^2 - 6x + 3$, then at what value of x will g(x) has a maximum value ?

(A)     -5

(B)     -3

(C)     3

(D)     7

(E)     12

19. Lee can write 5 reports in 6 hours, at this rate how many reports can he write in x days ?

(A)     20x

(B)     24x

(C)     30x

(D)     36x

(E)     72x

20. Matrix A has a dimension of 3 by 2, matrix B has a dimension of 2 by 4 and matrix C has a dimension of 1 by 2. Which of the following product of matrices is possible ?

(A)     **ACB**

(B)     **CAB**

(C)     **CBA**

(D)     **AC**

(E)     **CB**

21. A regular polygon with the interior angle of 160 degree each has how many sides ?

(A)     10

(B)     13

(C)     15

(D)     18

(E)     22

22. What is the volume (in $cm^3$) of a hemisphere with the radius of 1.2 cm ?

(A)     3.6

(B)     4.8

(C)     5.7

(D)     7.8

(E)     9.4

23. In a triangle ABC , AB = 12 , BC = 13 and AC = 16. What is the value of the angle CBA ?

(A)     39.7

(B)     45.8

(C)     79.5

(D)     86.7

(E)     112.5

SAT MATH LEVEL 1 Practice-Test

24. Which of the following shows the range of the trig function   $y = -10 \cdot \sin(x)$  ?

(A)     $-10 \le y \le 0$

(B)     $-10 < y < 10$

(C)     $0 \le y \le 10$

(D)     $1 < y < 10$

(E)     $-10 \le y \le 10$

25. Alex is buying a PS6 during a Christmas sales, which is originally cost $ 990. He is getting a discount of 10% plus another 10% on top if he purchases it online.  Which of the following expression can be used to calculate the amount Alex must pay if he is buying PS6 online?

(A)     $990 ( 1 - 0.2)$

(B)     $990 (0.1)(0.2)$

(C)     $990 (0.1)^2$

(D)     $990 ( 0.9)(0.1)$

(E)     $990 (0.9)^2$

26. If $f(x) = 5x - 3$ then  $f^{-1}(x) =$

(A)     $f^{-1}(x) = x + 0.6$

(B)     $f^{-1}(x) = 0.2x + 0.6$

(C)     $f^{-1}(x) = 5x + 3$

(D)     $f^{-1}(x) = 3x + 5$

(E)     $f^{-1}(x) = 0.2x - 3$

27. Which of the following is the remainder when $x^3 - 2x^2 + 4x - 12$  is divided by  $x - 4$

(A)     -124

(B)     -19

(C)     0

(D)     36

(E)     112

28. In an arithmetic sequence the first term is 5 and the ninth term is 45,

what is the second term of the sequence ?

(A)     6

(B)     8

(C)     10

(D)     13

(E)     15

29. What is the probability of rolling two six-sided dice and obtaining the sum as a square number ?

(A)     1/6

(B)     1/3

(C)     1/2

(D)     7/36

(E)     11/36

SAT MATH LEVEL 1 Practice-Test

30. If $f(3x) = 12x - 9$, then $f(x) =$

(A)     $4x - 9$

(B)     $4x - 3$

(C)     $\frac{(x-9)}{3}$

(D)     $36x - 9$

(E)     $36x - 27$

31. Which of the following number is the COUNTEREXAMPLE to the statement " All prime numbers are odd " ?

(A)     9

(B)     8

(C)     5

(D)     2

(E)     1

32. Evaluate $i^5 + i^6 - i^7 - i^8 =$

(A)     $2i$

(B)     $i - 2$

(C)     $i + 2$

(D)     $2i - 2$

(E)     $2i + 2$

33. If $\sin \theta = -\frac{4}{5}$ , then $\cos \theta =$

(A)     $-\frac{9}{5}$

(B)     $-\frac{7}{5}$

(C)     $-\frac{5}{3}$

(D)     $\frac{9}{5}$

(E)     $\frac{3}{5}$

34. $x^2 - 7x + 12 = (x - 3)^2$ is equivalent to

(A)     $3 - x = 0$

(B)     $x - 3 = 0$

(C)     $2x^2 - x + 3 = 0$

(D)     $2x^2 - 13x - 3 = 0$

(E)     $2x^2 + 3 = 0$

35. If a regular hexagon of side 8 cm has a circle inscribe inside, what is the area of this circle ?

(A)     $128 \pi$

(B)     $64 \pi$

(C)     $56 \pi$

(D)     $48 \pi$

(E)     $32 \pi$

36. A square has coordinates of (4,5) , (7,1) and (3, -2) will have the fourth coordinate of

(A)    ( 2, 0)

(B)    (-2, 2)

(C)    (0, -2)

(D)    (-2, 0 )

(E)    ( 0, 2 )

37. If $f(x) = \dfrac{x+11}{x^2-121}$ , and $f(a) = 0$ then $a = ?$

(A)    a = 11

(B)    a = 11 and a = 0

(C)    a = -11

(D)    a = -11 and a = 0

(E)    a = -11 and a = 11

38. $\left(-\dfrac{27}{125}\right)^{-1/3} =$

(A)    - 0.6

(B)    - 1.67

(C)    0.6

(D)    1.67

(E)    no real value

39. If $p + 2q < p - 3q$ then

(A)    p > 5

(B)    q < 5

(C)    pq < 1

(D)    q < 0

(E)    q > 0

40. The area between line $3y = -4x + 12$ , x-axis and y - axis is equal to

(A)    12

(B)    10

(C)    6

(D)    5

(E)    3

41. If the average of 5, 9 , 10 and x is 7 then which of the following is the value of x ?

(A)    4

(B)    5

(C)    7

(D)    8

(E)    11

42. Given that

$$f(x) = \begin{cases} 2x + 2 & -1 \le x < 0 \\ x^2 + 2 & 0 \le x < 6 \\ 38 & x \ge 6 \end{cases}$$

What is the range of f(x) ?

(A)         $-1 \le f(x) \le 38$

(B)         $0 \le f(x) \le 38$

(C)         $-1 < f(x) < 6$

(D)         $-6 \ge f(x) \ge 6$

(E)         $0 > f(x) \ge 6$

43. If **sinθ = 0.5**, then   $3\sin^2\theta + 3\cos^2\theta =$

(A)       $3 \sin(\theta) + 3 \cos(\theta)$

(B)       $3 \sin(\theta) - 3 \cos(\theta)$

(C)       $3 \sin(\theta) \cdot \cos(\theta)$

(D)       3

(E)       1.5

44. What is the greatest volume of the cylinder that can be drawn inside the box below?

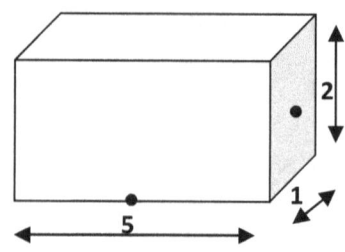

(A)       $1.0\,\pi$

(B)       $1.25\,\pi$

(C)       $2.25\,\pi$

(D)       $4.0\,\pi$

(E)       $6.25\,\pi$

45. If   $(\sqrt{5} + 1)(\sqrt{20} - 1) =$

(A)       $9 + \sqrt{5}$

(B)       $9 - \sqrt{5}$

(C)       $9\sqrt{5} - 1$

(D)       $3\sqrt{5} - 9$

(E)       $3\sqrt{5} + 9$

46. The graph of $x^2 + y^2 - 4 = 0$ is the graph of

(A)       an ellipse

(B)       a circle

(C)       a parabola

(D)       a hyperbola

(E)       a periodic

SAT MATH LEVEL 1 Practice-Test

47. Which of the following is the domain of $y = \sqrt{2x - 5}$ ?

(A)     All real number

(B)     $-2.5 \leq x \leq 2.5$

(C)     $0 \leq x \leq 2.5$

(D)     $-2.5 \leq x \leq 0$

(E)     $x \geq 2.5$

48. If $\dfrac{(n)!}{(n-1)!} = 60$ , then n =

(A)     61

(B)     60

(C)     59

(D)     57

(E)     55

49. If $f(x) = x^3 - 2x^2 + kx + 6$ is divisible by $x - 3$ then k = ?

(A)     -6

(B)     -5

(C)     3

(D)     5

(E)     6

50. The value $\sin(\dfrac{\pi}{2} - \theta) =$

(A)     $\cos\theta + \sin\theta$

(B)     $\cos\theta - \sin\theta$

(C)     $2\cos\theta\sin\theta$

(D)     $-\sin\theta$

(E)     $\cos\theta$

## END OF TEST 1

**1. E**

Method: **Solve**

$$U - V = -3$$

and $\quad U + V = -2$

$$U^2 - V^2 = (U - V)(U + V)$$

$$U^2 - V^2 = (-3)(-2)$$

$$U^2 - V^2 = \underline{6}$$

**2. C**

Method: **Plug into the formula**

$$\text{slope} = \frac{y_2 - y_1}{x_2 - x_1} = \frac{5 - 5}{2 + 1} = \frac{0}{3} = \underline{0}$$

Slope = 0

**3. D**

Method: **Evaluate**

Sum to fifth term = $3^0 + 3^{-1} + 3^{-2} + 3^{-3} + 3^{-4}$

Sum to fifth term = **121/81**

**4. E**

Method: **Simplify**

$$a = \sqrt[3]{4 - 12}$$

$$a = \sqrt[3]{-8}$$

$$\underline{a = -2}$$

**5. C**

Method: **Solve**

$$C = 2\pi R \rightarrow 18\pi = 2\pi R \rightarrow R = 9$$

$$A = \pi R^2 \rightarrow A = \pi (9)^2 = \mathbf{81\pi}$$

**6. C**

Method: **List all the slopes out**

We know that y = 2x -12 has a **slope of 2**

find the choice with the slope of 2

| Choice | Equation | Slope |
|--------|----------|-------|
| A | y = 0.5x + 12 | 0.5 |
| B | y = -2x + 12 | -2 |
| C | x = 0.5y +12 | 2 |
| D | x = 0.2y +12 | 5 |
| E | x = y | 1 |

Choice C gives the slope of 2,

so line x = 0.5y + 12 is parallel to line y = 2x -12

**7. D**

Method: **Evaluate**

$$b(0) = 5^0 = 1$$

$$a(b(0)) = a(1) = 5^1$$

$$a(b(0)) = \mathbf{5}$$

**8. D**

Method: **Graph and Solve**

Draw the graph of $y = -x^2 + 16$

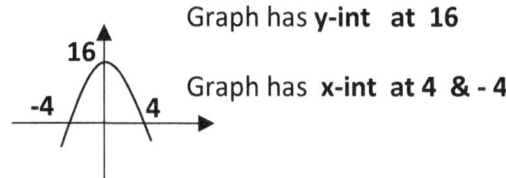

Graph has **y-int** at **16**

Graph has **x-int** at **4 & -4**

The x-intercepts are at -4 and 4

**9. B**

Method: **Graph**

Graph of a function means the graph should be one to one relation or many to one relation. We can use a vertical line test to see if the graph drawn has more than one value of y touching the vertical line or not, if there is then it is not a function.

|   | Equation | Vertical line test |
|---|----------|--------------------|
| A | $x = 1$ | Touch the graph infinitely |
| B | $y = 1$ | Touch the graph once |
| C | $x = 0$ | Touch the graph infinitely |
| D | $y = \pm \sqrt{x}$ | Touch the graph twice |
| E | $xy^{-4} = 1$ | Touch the graph twice |

**10. A**

Method: **Simplify the drawing**

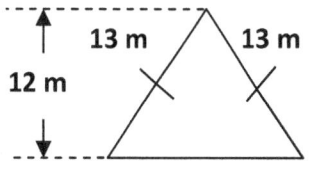

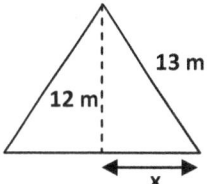

Using Pythagoras → $12^2 + x^2 = 13^2$

$x = 5$ ← double this = base

The base of the triangle is 5 x 2 = 10

So , perimeter = 13 + 13 + 10 = 33

**11. C**

Method: **Solve**

$\log_a X = b + d$ then $X = $ ??

$\log_a X = \log_a 9 + \log_a 2$

$\log_a X = \log_a (9 \times 2)$ ← applying addition rule

$\log_a X = \log_a (18)$ ← equating the inside

$X = \underline{\underline{18}}$

## 12. D

Method: **Evaluate**

Area $= \pi R^2$

|  | Radius (R) | Volume (V) |
|---|---|---|
| Before | 1 | $1\,\pi$ |
| After double | 2 | $4\,\pi$ |

$$\text{\% change} = \frac{V_{After} - V_{Before}}{V_{before}} \times 100\%$$

$$\text{\%change} = \frac{4\pi - \pi}{\pi} \times 100\%$$

$$\text{\%change} = \mathbf{\underline{300\%}}$$

## 13. B

Method: **Evaluate**

Profit = Revenue - Cost

Profit = 480(20) - 250(20)

Profit = **4600**

## 14. D

Method: **Evaluate and Solve**

$$2x - 3 = (x - 1)^2$$

$$2x - 3 = x^2 - 2x + 1$$

$$0 = x^2 - 4x + 4$$

$$0 = (x - 2)(x - 2)$$

$$x = 2 \quad only$$

## 15. C

Method: **Plug into the formula**

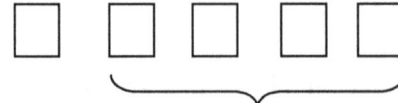

President        The rest of the class

$$1 \qquad x \qquad {}_4P_4$$

$$= \mathbf{\underline{24}}$$

## 16. D

Method: **Plug into the formula**

$$2 \sin(2x) - 3$$

$$\text{Period} = \frac{2\pi}{2} = \pi$$

## 17. B

Method: **Evaluate**

$$y = \frac{e^x}{x-2} = \frac{e^2}{2-2} = \frac{e^2}{0} = \text{Undefined}$$

We do not want the denominator to be zero. **Therefore x cannot be equal to 2.**

18. **B**

Method: **Plug-in the formula**

$g(x) = -x^2 - 6x + 3$

Vertex → $x = \frac{-b}{2a}$

$x = \frac{-(-6)}{2(-1)}$

$x = -3$

**Therefore minimum occurs at x equal to -3**

19. **A**

Method: **Take ratio**

$\frac{Time}{Reports} \rightarrow \frac{6\ hours}{5\ reports} = \frac{X\ days}{??Reports}$

$\frac{6\ hours}{5\ reports} = \frac{X \times 24\ hours}{??Reports}$

$??\ Reports = \frac{24X \times 5}{6} = \underline{\mathbf{20x}}$

**Lee can write 20x reports in x days**

20. **E**

Method: **Evaluate**

We know that for a matrix to multiply the column of the first matrix must equal to the row of the second matrix. The only possibility here is at **CB.**

21. **D**

Method: **Evaluate using formula**

**We know that each interior angle of a regular polygon can be found using formula:**

$\frac{(n-2)180}{n} = 160$ , n is number of sides

180n - 360 = 160n

**n = 18**

22. **A**

Method: **Plug into the formula**

radius = 1.2 cm

Volume of hemisphere = 2/3 π r³ = 2/3 π· (1.2)³

Volume of sphere = **3.6**

23. **C**

Method: **Plug into the formula**

**USE** cosine rule

$\cos\theta = \frac{c^2 - a^2 - b^2}{-2ab}$

$\cos(CBA) = \frac{16^2 - 12^2 - 13^2}{-2(12)(13)}$

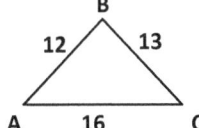

**CBA = 79.5°**

**24. B**

Method: **Graph**

$y = -10\sin(x)$

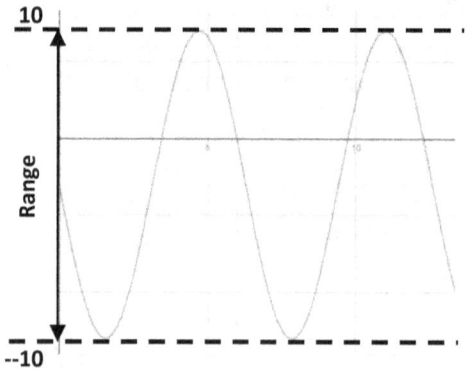

Range is between $-10 \le y \le 10$

**25. E**

Method: **Plug into the formula**

| discount 10% on 990 |
| paid amount → 990 x 0.9 |

| discount another 10% on 990 x 0.9 |
| final amount → 990 x 0.9 x 0.9 = **$ 990(0.9)²** |

**26. B**

Method: **Solve and Simplify**

$$y = 5x - 3$$

swap x and y → $x = 5y - 3$

make 'y' the subject

$$5y = x + 3$$

$$y = (x + 3)/5$$

$$y = 0.2x + 0.6$$

$$f^{-1}(x) = 0.2x + 0.6$$

**27. D**

Method: **Evaluate**

using remainder theorem

$$R(x) = x^3 - 2x^2 + 4x - 12$$

$$R(4) = 4^3 - 2(4)^2 + 4(4) - 12$$

$$R(4) = \underline{\mathbf{36}}$$

**28. C**

Method: **Plug into the formula**

$$U_n = U_1 + (n-1)d$$

$$U_9 = U_1 + 8d \rightarrow 45 = 5 + 8d \rightarrow d = 5$$

$$U_2 = U_1 + d \rightarrow U_2 = 5 + 5 \rightarrow U_2 = 10$$

**The second term is 10**

**29. D**

Method: **Create table of possiblities**

|   | 1 | 2 | 3 | 4 | 5 | 6 |
|---|---|---|---|---|---|---|
| 1 | 2 | 3 | 4 | 5 | 6 | 7 |
| 2 | 3 | 4 | 5 | 6 | 7 | 8 |
| 3 | 4 | 5 | 6 | 7 | 8 | 9 |
| 4 | 5 | 6 | 7 | 8 | 9 | 10 |
| 5 | 6 | 7 | 8 | 9 | 10 | 11 |
| 6 | 7 | 8 | 9 | 10 | 11 | 12 |

The square number appeared seven times out of 36 possibilities.

$$P(\text{ sum as a square number }) = \frac{7}{36}$$

## 30. A

Method: **Simplify**

$$f(3x) = 12x - 9$$

$$f(3(\tfrac{x}{3})) = 12(\tfrac{x}{3}) - 9$$

$$f(x) = 4x - 9$$

## 31. D

Method: **Evaluate**

" All prime numbers are odd "

We know that there is an even number **, 2 ,** that is a prime number also this number will contradict the statement above.

## 32. D

Method: **Evaluate**

We know that **:** $i^4 = 1$ **;** $i^3 = -i$ **;** $i^2 = -1$

$$i^5 = i^4 \cdot i^1 = i \qquad\qquad i^6 = i^4 \cdot i^2 = -1$$

$$i^7 = i^4 \cdot i^3 = -i \qquad\qquad i^8 = i^4 \cdot i^4 = 1$$

Plug all the values back in:

$$i^5 + i^6 - i^7 - i^8$$

$$= \quad i + (-1) - (-i) - 1 = \mathbf{\mathit{2i - 2}}$$

## 33. E

Method: **Plug into the formula**

$$\sin\theta = \frac{opp}{hyp} = \frac{-4}{5}$$

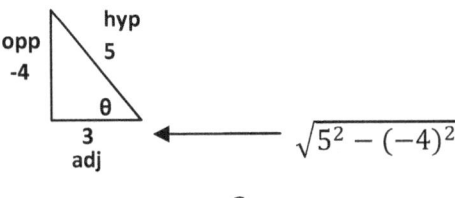

$$\cos\theta = \frac{adj}{hyp} = \frac{3}{5}$$

## 34. A

Method: **Expand and Simplify**

$$x^2 - 7x + 12 = (x - 3)^2$$

$$x^2 - 7x + 12 = x^2 - 6x + 9$$

$$3 - x = 0$$

## 35. D

Method: **Solve by drawing**

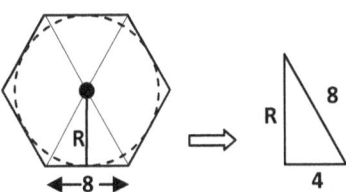

Pythagoras the small triangle

$$R^2 + 4^2 = 8^2 \;\rightarrow R^2 = 48$$

Area $= \pi R^2 = \pi (48)$

**Area = 48π**

## 36. **E**

Method: **Draw out and Evaluate**

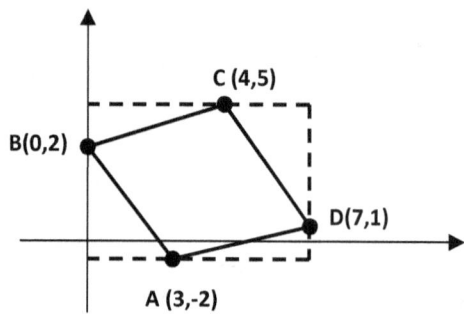

We know that the slope of the line AB and CD should be equal, since they are parallel.

Slope AB        Slope CD

$$\frac{y-(-2)}{x-3} = \frac{5-1}{4-7}$$

$$\frac{y+2}{x-3} = \frac{4}{-3}$$

Equating the top and bottom

$y + 2 = 4 \rightarrow \quad y = 2$

$x - 3 = -3 \rightarrow \quad x = 0$

**coordinate is (0,2)**


## 37. **C**

Method: **Evaluate**

$$f(x) = \frac{x+11}{x^2-121} \text{ , and } f(a) = 0$$

$$f(a) = \frac{a+11}{a^2-121} \rightarrow 0 = \frac{a+11}{a^2-121}$$

$$a + 11 = 0$$

$$a = -11$$


## 38. **B**

Method: **Evaluate**

$$\left(-\frac{27}{125}\right)^{-1/3} = \left(-\frac{125}{27}\right)^{1/3}$$

$$= \left(-\frac{5^3}{3^3}\right)^{1/3}$$

$$= -\frac{5}{3} = -1.67$$


## 39. **D**

Method: **Simplify**

$$p + 2q \quad < \quad p - 3q$$

$$2q + 3q \quad < \quad p - p$$

$$5q \quad < 0$$

$$\mathbf{q} \quad \mathbf{< 0}$$


## 40. **C**

Method: **Graph and Evaluate**

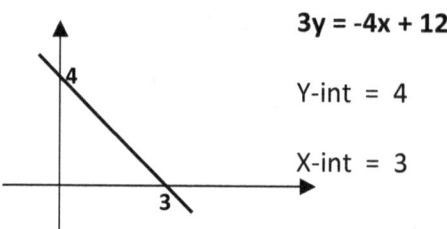

$3y = -4x + 12$

Y-int = 4

X-int = 3

We know that the figure enclosed will be a triangular shape, so we use formula

Area $= \frac{1}{2}$ base x height $= \frac{1}{2}$ x 3 x 4

**Area = 6 unit$^2$**

## 41. A

Method: **Solve using formula**

$$\frac{5+9+10+x}{4} = 7$$

$$24 + x = 28$$

$$x = 4$$

## 42. B

Method: **Graph**

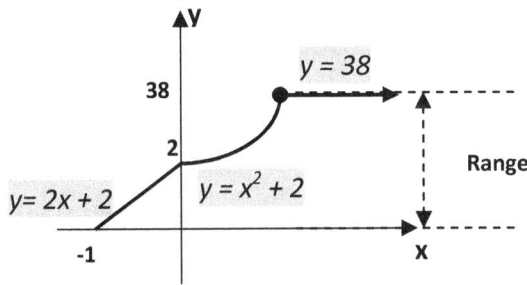

Range: $0 \leq y \leq 38$

## 43. D

Method: **Plug into the formula**

$3\sin^2\theta + 3\cos^2\theta = 3(\sin^2\theta + \cos^2\theta) = 3(1)$

$= 3$

## 44. A

Method: **Solve by drawing**

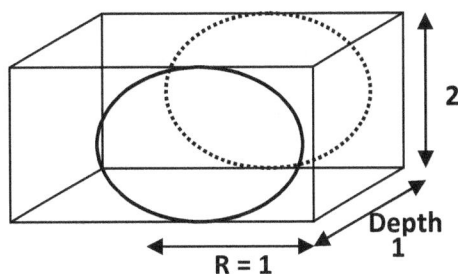

Volume $= \pi R^2 \times$ depth

Volume $= \pi (1)^2 \times 1$

## 45. A

Method: **Simplify by expansion**

$$(\sqrt{5} + 1)(\sqrt{20} - 1)$$

$$= \sqrt{100} - \sqrt{5} + \sqrt{20} - 1$$

$$= 10 - \sqrt{5} + 2\sqrt{5} - 1$$

$$= 9 + \sqrt{5}$$

## 46. B

Method: **Graph**

$$x^2 + y^2 = 4 \quad \text{equation of circle}$$

$$x^2 + y^2 = R^2$$

The graph is the graph of circle

47. **E**

Method: **Evaluate**

$y = \sqrt{2x - 5}$ → We know that inside the root the value cannot be negative so we say

**2x - 5 ≥ 0** → **2x ≥ 5** → **x ≥ 2.5**

**Domain : x ≥ 2.5**

48. **B**

Method: **Simplify**

$$\frac{n(n-1)!}{(n-1)!} = 60$$

n = 60

49. **B**

Method:

$(x^3 - 2x^2 + kx + 6) \div (x-3)$

We know that x =3 is the root of f(x)

We apply remainder theorem here

f(3) = 0

$(3)^3 - 2(3)^2 + k(3) + 6 = 0$

15 + 3k = 0

k = **-5**

50. **E**

Method: **Plug into the formula**

$\sin(\frac{\pi}{2} - \theta) = \sin(\frac{\pi}{2}) \cos\theta - \cos(\frac{\pi}{2}) \sin\theta$

= (1) cos θ - (0) sin θ

= **cos θ**

| Raw Score | Conversion |
|-----------|------------|
| 44 - 50 | 800 |
| 39 - 43 | 750 - 790 |
| 36 - 38 | 720 - 740 |
| 33 - 35 | 690 - 710 |
| 29 - 32 | 650 - 680 |
| 25 - 28 | 590 - 640 |
| 20 - 24 | 540 - 580 |
| 25 - 29 | 510 - 530 |
| 20 - 24 | 450 - 500 |
| 15 - 19 | 400 - 440 |

Raw Score　　=　　Correct Answers　-　0.25 Wrong Answers

$$\boxed{\phantom{xxx}} \;=\; \boxed{\phantom{xxx}} \;-\; 0.25 \times \boxed{\phantom{xxx}}$$

1. If $x^2 - 3 = 6$ and $y^2 - 4 = 12$ then $x \cdot y =$

(A)    1

(B)    3

(C)    4

(D)    7

(E)    12

2. What is the distance between point **(3, 2)** and **( -3, -6)** ?

(A)    10

(B)    9

(C)    4

(D)    3

(E)    1.33

3. If $f(x) = 3 \cdot x - 12$ , what is the x-intercept of f(x)?

(A)    -12

(B)    -4

(C)    -3

(D)    4

(E)    12

4.    $\dfrac{1}{2\sqrt{3}-1} =$

(A)    $\dfrac{6+\sqrt{3}}{11}$

(B)    $\dfrac{6-\sqrt{3}}{11}$

(C)    $\dfrac{1+2\sqrt{3}}{11}$

(D)    $\dfrac{1+\sqrt{3}}{12}$

(E)    $\dfrac{2+2\sqrt{3}}{3}$

5. If $\sec \theta = \dfrac{13}{12}$ , then $\tan \theta =$

(A)    $\dfrac{12}{13}$

(B)    $\dfrac{13}{15}$

(C)    $\dfrac{13}{5}$

(D)    $\dfrac{12}{15}$

(E)    $\dfrac{5}{12}$

6. If $a^{2x} + a^x = a^5(a^5+1)$ then $x =$

(A)    -5

(B)    -3

(C)    0

(D)    3

(E)    5

7. If there were 5 boys and 4 girls in a Sudoku club then, how many ways can 3 boys and 3 girls be chosen to compete in world-wide tournament ?

(A)     40

(B)     60

(C)     120

(D)     720

(E)     1440

8. The sequence **2, 6, 1, 7, 0, ….** has the 8th term equal to

(A)     -8

(B)     -1

(C)     0

(D)     8

(E)     9

9. If $x^3 = x$ , then the solution must consist of

(A)     0 only

(B)     1 only

(C)     -1 only

(D)     0 and 1

(E)     -1, 0 and 1

10. The graph **y = x² -4x + 7** has the vertex value at **x =**

(A)     -2

(B)     -0.5

(C)     0

(D)     0.5

(E)     2

11. The diagonal of a square is 20 cm long, what is the radius of the largest circle that can fit inside the square ?

(A)     $2\sqrt{5}$

(B)     $5\sqrt{2}$

(C)     $3\sqrt{5}$

(D)     $5\sqrt{3}$

(E)     $2\sqrt{3}$

12.  $\sqrt[3]{-\dfrac{54}{16}}$  =

(A)     - 0.544

(B)     -0.984

(C)     - 1.50

(D)     0.544

(E)     1.50

SAT MATH LEVEL 1 Practice-Test

13. The maximum value of **4 - 2sin(x - π)** =

(A)     0

(B)     2

(C)     4

(D)     6

(E)     8

14. If $g(x) = e^{2x+5}$ then $g^{-1}(1) =$

(A)     -5

(B)     -2.5

(C)     0

(D)     0.4

(E)     1

15. If **sin (x) = 0.5** then **x =**

(A)     50

(B)     45

(C)     60

(D)     120

(E)     150

16. What is the equation of axis of symmetry of the graph with equation $y = (x-5)^2 + 3$ ?

(A)     x = -3

(B)     x = 3

(C)     x = 5

(D)     y = 5

(E)     y = 3

17. Arithmetic sequence **5, a , b, ....** has the seventh term equal to 20.  What is the value of **a + b** ?

(A)     20.5

(B)     17.5

(C)     15

(D)     10

(E)     12.5

18. If $g(x) = \log_2 (x + 1) - 2$ and $g(a) = 0$

then a =

(A)     **4**

(B)     **3**

(C)     **2**

(D)     **1**

(E)     **0**

19. To pass math exam Delan must get the average mark of more than 64 in all eight tests. His average now is 60 for the seven tests. What is the minimum mark he must get on his eight test ?

(A)     80

(B)     84

(C)     92

(D)     93

(E)     96

20. The range of a function $f(x) = \frac{5}{x-3}$ is

(A)             $g(x) < 0$

(B)             $g(x) \geq 0$

(C)     $-\infty < g(x) < 5$ and $5 < g(x) < \infty$

(D)     $-\infty < g(x) < 0$ and $0 < g(x) < \infty$

(E)             All real number

21. If $f(3) = 0$ and $f(9) = 12$ and f(x) is a linear function then what is 'a' if $f(a) = 20$ ?

(A)     14

(B)     13

(C)     12

(D)     11

(E)     10

22. The period of graph $\tan(3\pi \cdot x - 4) + 6$ is

(A)     $3\pi$

(B)     $2\pi$

(C)     $\pi/3$

(D)     $3/2$

(E)     $1/3$

23. The equation of the circle $x^2 - 6x + y^2 + 8y = -9$ has the **center** and **radius** equal to

|       | center  | radius |
|-------|---------|--------|
| (A)   | (3,-4)  | 4      |
| (B)   | (-3,4)  | 4      |
| (C)   | (-3,4)  | 3      |
| (D)   | (-3,-4) | 3      |
| (E)   | (3,4)   | 2      |

SAT MATH LEVEL 1 Practice-Test

24. Triangle ABC is similar to triangle XYZ, if side AB = 3 , BC = 4 and XY = 4.5 then XZ =

(A)   5.0

(B)   6.5

(C)   7.0

(D)   7.5

(E)   8.0

25. The probability that it will rain today is 0.4 and the probability that there will be traffic on any given day is 0.7 , what is the probability that there will be no rain and no traffic today ?

(A)   0.90

(B)   0.70

(C)   0.63

(D)   0.28

(E)   0.18

26. If $\frac{2}{3}$ is $\frac{1}{2}$ of $\frac{5}{6}$ of a certain number, then what is the number ?

(A)   $\frac{5}{8}$

(B)   $\frac{8}{5}$

(C)   $\frac{8}{15}$

(D)   $\frac{8}{3}$

(E)   $\frac{3}{8}$

27. What is the shortest distance between the point ( -2, -3) and line x = 4 ?

(A)   2.0

(B)   4.5

(C)   6.0

(D)   7.0

(E)   8.5

28. If **a** and **b** are positive integer and **ab = 6** then which of the following cannot be $\frac{a}{b}$ ?

(A)   0.167

(B)   0.667

(C)   1.5

(D)   1.667

(E)   6.0

29. $\dfrac{1-\frac{2}{x+1}}{1+\frac{3}{2x-5}}$ =

(A)   $\frac{2x-5}{2x-1}$

(B)   $\frac{2x-5}{2x+2}$

(C)   $\frac{2x+5}{x-1}$

(D)   $\frac{2x+5}{2x+2}$

(E)   $\frac{2x+5}{2x+1}$

SAT MATH LEVEL 1 Practice-Test

30. Let $a@b = \frac{ab^2}{a^b}$ , what is (2 @4) - (3 @ 2) ?

(A)     2/3

(B)     3/2

(C)     4/9

(D)     -4/9

(E)     -3/2

31. If point U and V lie on the circular cylinder with radius of 3 and height of 12, then what is the maximum possible straight line distance between point U and V ?

(A)     12.3

(B)     12.8

(C)     13.4

(D)     14.6

(E)     18.0

32. If a is a positive integer, for which value of b is |b-a| be greatest?

(A)     3

(B)     5

(C)     0

(D)     -5

(E)     -3

33. In triangle XYZ if angle y is a right angle then which of the following is equivalent to sin Z  ?

(A)     $\frac{XY}{ZY}$

(B)     $\frac{XY}{ZX}$

(C)     $\frac{XZ}{ZY}$

(D)     $\frac{XZ}{ZX}$

(E)     $\frac{ZY}{XY}$

SAT MATH LEVEL 1 Practice-Test

|  | Day 1 | Day 2 | Day 3 |
|---|---|---|---|
| Samsung | 213 | 223 | 201 |
| Apple | 294 | 334 | 324 |
| Sony | 112 | 145 | 109 |

34. The table above shows the number of tablet sold during a three-day period. The prices of tablet brand Samsung , Apple ,
and Sony were $300,$ 305 , and $280 respectively. Which of the following matrix representations gives the income, in dollars, received of these three days?

(A)
$$\begin{vmatrix} 213 & 223 & 201 \\ 294 & 334 & 324 \\ 112 & 145 & 109 \end{vmatrix} \begin{vmatrix} 300 & 305 & 280 \end{vmatrix}$$

(B)
$$\begin{vmatrix} 213 & 223 & 201 \\ 294 & 334 & 324 \\ 112 & 145 & 109 \end{vmatrix} \begin{vmatrix} 300 \\ 305 \\ 280 \end{vmatrix}$$

(C)
$$\begin{vmatrix} 280 & 305 & 300 \end{vmatrix} \begin{vmatrix} 213 & 223 & 201 \\ 294 & 334 & 324 \\ 112 & 145 & 109 \end{vmatrix}$$

(D)
$$\begin{vmatrix} 300 \\ 305 \\ 280 \end{vmatrix} \begin{vmatrix} 213 & 223 & 201 \\ 294 & 334 & 324 \\ 112 & 145 & 109 \end{vmatrix}$$

(E)
$$\begin{vmatrix} 300 & 305 & 280 \end{vmatrix} \begin{vmatrix} 213 & 223 & 201 \\ 294 & 334 & 324 \\ 112 & 145 & 109 \end{vmatrix}$$

35. The class of eight grader took a test, the mean score was 80, the median was 76 and the standard deviation of the scores was 6. The homeroom teacher decided to add 3 points to each student's scores as a bonus. Which of the following statement is valid ?

    I.      the new mean is  83

    II.    the new median is 79

    III.   the new standard deviation is 9

(A)    I only

(B)    II only

(C)    I and II only

(D)    I , II and III

(E)    none of the above

36. Mike was driving his La Ferrari with the speed of 150 mph on the autobahn for 2.5 hours then he reduces his speed as he enters Munich and travel further more. If the entire journey took him four hours to cover a distance of 495 miles then at what speed was he driving as he enters Munich ?

(A)    70    mph

(B)    80    mph

(C)    95    mph

(D)    113    mph

(E)    124    mph

37. Point U has a coordinate of (3,0) and the pentagon below has all side of length 5 units. What is the area of triangle UVW ?

(A)    12.1

(B)    10.4

(C)    9.63

(D)    8.27

(E)    7.71

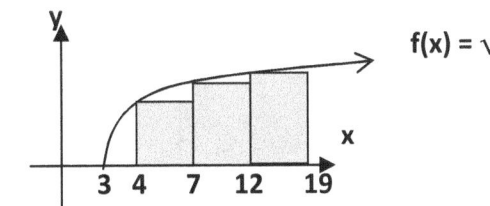

38. What is the sum of the shaded area below the graph f(x) ?

$$f(x) = \sqrt{x - 3}$$

(A)    19

(B)    21

(C)    27

(D)    34

(E)    37

39. If $f(x) = \frac{2x+1}{1-x}$ then $f^{-1}(x) =$

(A)    $\frac{2x-1}{1-x}$

(B)    $\frac{x-1}{2x+1}$

(C)    $\frac{x-1}{2+x}$

(D)    $\frac{x+1}{2-x}$

(E)    $\frac{x+1}{1-2x}$

40. What is the middle term of the expansion of $(2x - 3)^4$ ?

(A)    $-128\,x^3$

(B)    $216\,x^3$

(C)    $-96\,x^3$

(D)    $216\,x^2$

(E)    $96\,x^2$

41. A coin is tossed three times, what is the probability that at least two tails appear ?

(A)    3/8

(B)    1/2

(C)    1/3

(D)    1/6

(E)    1/8

42. The solution set of $5y - 3x > 0$ lies in which quadrants ?

(A)    I and II only

(B)    I and III only

(C)    II and IV only

(D)    I, II and III only

(E)    II , III and IV only

43. A hexagon is inscribe inside a circle with a radius of 5 cm. What is the area of this hexagon?

(A)     10.8

(B)     15.8

(C)     34.6

(D)     52.3

(E)     65.0

44. What is the domain of $f(x) = \sqrt{5 - x}$ ?

(A)     $x \geq 0$

(B)     $x \leq 0$

(C)     $x \geq -5$

(D)     $x \leq -5$

(E)     $x \leq 5$

45. The surface area of the cube is equal to its volume, what is the volume of the

largest sphere that can fit inside the cube ?

(A)     $36\,\pi$

(B)     $72\,\pi$

(C)     $108\,\pi$

(D)     $144\,\pi$

(E)     $288\,\pi$

46. What is the sum of the sequence  2, 1, 0.5, , ...... to infinite term ?

(A)     3.0

(B)     4.0

(C)     4.5

(D)     5.0

(E)     5.5

47. If $\ln(x + 3) = 2 \ln(3)$ , then  $x =$

(A)     0

(B)     2

(C)     4

(D)     6

(E)     8

48. $(\sin \theta - \cos \theta)^2 - 1 =$

(A)     $\cos 2\theta$

(B)     $\sin 2\theta$

(C)     $1 - \sin \theta - \cos \theta$

(D)     $-\sin 2\theta$

(E)     $-\cos 2\theta$

49. If 8 people in a class like chocolate flavor ice-cream , 6 people like both chocolate and vanilla flavor and 2 dislike both flavor. There were 20 people in a class. How many people like only vanilla flavor ice-cream ?

(A)      12

(B)      10

(C)      8

(D)      6

(E)      4

50. If  the graph of $y = x^2 + kx + 81$  has two equal root then which of the following is the value of k ?  ( k > 0)

(A)      6

(B)      9

(C)      18

(D)      20

(E)      24

**END OF TEST 2**

**1. E**

Method: **Solve**

$x^2 - 3 = 6 \rightarrow x^2 = 9 \rightarrow x = \pm 3$

$y^2 - 4 = 12 \rightarrow y^2 = 16 \rightarrow y = \pm 4$

$\qquad x \cdot y = 3 \cdot 4 = \mathbf{12}$

or $\qquad x \cdot y = -3 \cdot -4 = \mathbf{12}$

**2. A**

Method: **Plug into the formula**

$\text{dist} = \sqrt{(y_2 - y_1)^2 + (x_2 - x_1)^2}$

$\text{dist} = \sqrt{(-6 - 2)^2 + (-3 - 3)^2}$

$\text{dist} = \sqrt{100}$

$\mathbf{dist} = \underline{\underline{\mathbf{10}}}$

**3. D**

Method: **Evaluate**

x-intercept $\rightarrow$ y = 0, x = ?

$\qquad f(x) = 3x - 12$

$\qquad 0 = 3x - 12$

$\qquad x = \underline{\underline{4}}$

**4. C**

Method: **Simplify by multiplying the conjugate**

$$\frac{1}{2\sqrt{3}-1} \cdot \frac{2\sqrt{3}+1}{2\sqrt{3}+1}$$

$$= \frac{1 + 2\sqrt{3}}{(4\times 3) - 1} = \frac{1 + 2\sqrt{3}}{11}$$

**5. E**

Method: **Simplify using trig**

$\sec \theta = \dfrac{13}{12} = \dfrac{hyp}{adj}$

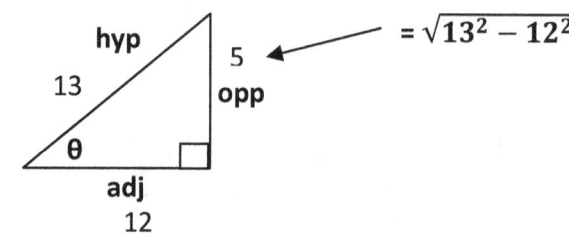

$\qquad \tan \theta = \dfrac{opp}{adj} = \dfrac{5}{12}$

**6. E**

Method: **Solve by factorizing**

$\qquad a^{2x} + a^x = a^5(a^5+1)$

$a^x (a^x + 1) = a^5(a^5+1)$

BASE ARE SAME $\rightarrow$ SO POWER ARE EQUAL

therefore, **x = 5**

**7. A**

Method: **Plug into the formula**

5 Boys  choose 3   x   4 Girls choose 3

$$^5C_3 \quad \times \quad ^4C_3$$

$$= \ 10 \ \times \ 4 \ = \ \underline{\textbf{40}}$$

**8. E**

Method: **Plug into the formula**

We know that the sequence is the alteration increment of +4, -5, +6, -7, +8, ...

| Term | 1 | 2 | 3 | 4 | 5 | 6 | 7 | 8 |
|------|----|----|----|----|----|----|----|----|
| Seq | 2 | 6 | 1 | 7 | 0 | 8 | -1 | **9** |
| Inc | +4 | -5 | +6 | -7 | +8 | -9 | +10 | |

**9. E**

Method: **Factorize and Solve**

$$x^3 = x \ \rightarrow \ x^3 - x = 0$$

$$x(x^2 - 1) = 0$$

$$x = 0 \ , \ x = \pm 1$$

**The values that will make them equal would be -1 , 0 and 1**

**10. E**

Method: **Plug in to the formula**

$$y = x^2 - 4x + 7 \ \rightarrow \qquad b = -4 \, , a = 1$$

vertex formula $\rightarrow x = \dfrac{-b}{2a} \ = \dfrac{-(-4)}{2(1)} \ = \ 2$

**11. B**

Method: **Draw out and Evaluate**

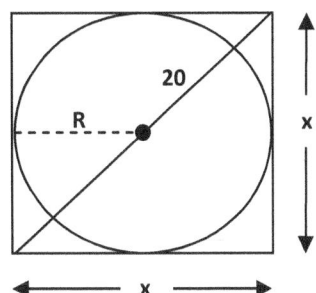

using Pythagoras theorem:  $x^2 + x^2 = 20^2$

$$2x^2 = 400$$

$$x = \sqrt{200} = 10\sqrt{2}$$

We know that R = 0.5x $\rightarrow$  R = 0.5 $(10\sqrt{2})$ = **$5\sqrt{2}$**

**12. C**

Method: **Evaluate**

$$\sqrt[3]{-\frac{54}{16}} \ = \ \sqrt[3]{-\frac{27}{8}} \ = \left[\frac{-3^3}{2^3}\right]^{1/3}$$

$$= \frac{-3}{2} \ = \ -\textbf{1.5}$$

**13. D**

Method: **Evaluate**

$$y \ = \ 4 - 2\sin(x - \pi)$$

maximum value occurs
when sin(x-π) = -1

$$y \ = \ 4 - 2(-1) \ = \ 4 + 2$$

$$y_{max} = \textbf{6}$$

**14. B**

Method: **Evaluate and Solve**

$g^{-1}(1) = ?$ → means if y =1 then x = ?

$g(x) = \boxed{y = e^{2x+5}}$

$1 = e^{2x+5}$ → $e^0 = e^{2x+5}$

$0 = 2x + 5$ → $\underline{x = -2.5}$

**15. E**

Method: **Solve**

$\sin(x) = 0.5$ → $x = \sin^{-1}(0.5)$

$x = 30$ and $\boxed{x = 150}$

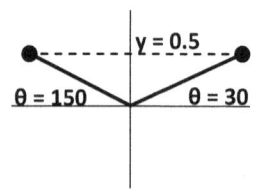

**16. C**

Method: **Graph**

$y = (x-5)^2 + 3$

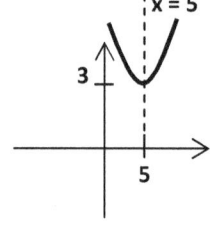

The line of symmetry has an equation of **x = 5**

**17. B**

Method: **Evaluate using formula**

arithmetic sequence formula

$$U_n = U_1 + (n-1)d$$

$U_7 = U_1 + (7-1)\,d$

$20 = 5 + 6\,d$

$d = 2.5$

Now we can find value of a and b

$a = 5 + 2.5 = 7.5$

$b = 7.5 + 2.5 = 10$

$a + b = 7.5 + 10 = \underline{17.5}$

**18. B**

Method: **Simplify**

$g(x) = \log_2(x + 1) - 2$

$g(a) = 0$ → $0 = \log_2(a + 1) - 2$

$2 = \log_2(a + 1)$

$2^2 = a + 1$

$4 - 1 = a$

$a = 3$

## 19. D

Method: **Solve**

Average on seven tests equals 60:

$$\frac{Test_1 + Test_2 + \dots Test_7}{7} = 60$$

$$Test_1 + Test_2 + \dots Test_7 = 420$$

Average on eight tests more than 64:

$$\frac{Test_1 + Test_2 + \dots Test_7 + Test_8}{8} > 64$$

$$Test_1 + Test_2 + \dots Test_7 + Test_8 > 512$$

$$420 + Test_8 > 512$$

$$Test_8 > 92$$

Therefore, he must obtain **93** marks in his eight test to pass the exam.

## 20. D

Method: **Graph**

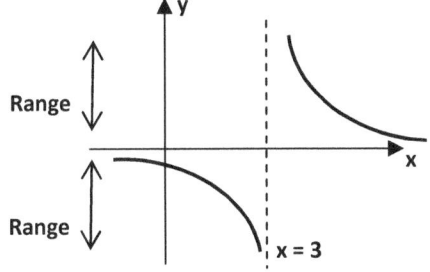

x = 3

**Range:** $-\infty < y < 0$ and $0 < y < \infty$

## 21. B

Method: **Plug into the formula**

**f(3) = 0** → $x_1 = 3$ and $y_1 = 0$

**f(9) = 12** → $x_2 = 9$ and $y_2 = 12$

$$slope = \frac{12-0}{9-3} = 2$$

**f(a) = 20** → x = a and y = 20

using slope equation with point (3,0)

$$slope = \frac{20-0}{a-3}$$

$$2 = \frac{20-0}{a-3}$$

$$2a - 6 = 20$$

$$a = \underline{13}$$

## 22. E

Method: **Plug into the formula**

$$tan(3\pi \cdot x - 4) + 6$$

$$Period = \frac{\pi}{3\pi} = \frac{1}{3}$$

## 23. A

Method: **Simplify and Plug into the formula**

Complete the square

$$x^2 - 6x + y^2 + 8y = -9$$

$$(x - 3)^2 + (y + 4)^2 = 16 \leftarrow (x - h)^2 + (y - k)^2 = r^2$$

$$(h, k) = (3, -4)$$

$$r = 4$$

## 24. D

Method: **Draw out and Evaluate**

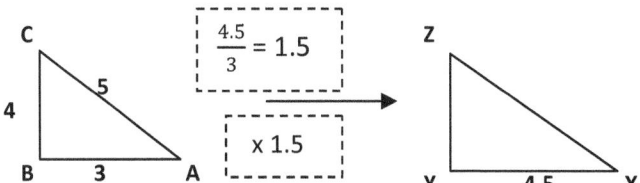

We can find AC by using Pythagoras

$$AC = \sqrt{3^2 + 4^2} = 5$$

We know that XYZ is the enlargement by factor of 1.5, so XZ = 1.5 of AC → XZ = 1.5 x 5 = **7.5**

**25. E**

Method: **Evaluate**

P(no rain and no traffic)

= P(no rain)  x P(no traffic)

= [1-0.4]  x  [1-0.7]

= 0.6 x 0.3

=  **0.18**

**26. B**

Method: **Evaluate**

$$\frac{2}{3} = \frac{1}{2} \cdot \frac{5}{6} \cdot x$$

$$x = \frac{2}{3} \cdot \frac{2}{1} \cdot \frac{6}{5}$$

$$x = \frac{8}{5}$$

**27. C**

Method: **Draw out the graph**

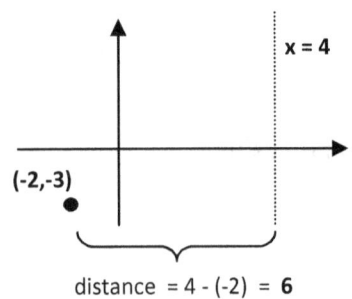

distance = 4 - (-2) = **6**

**28. D**

Method: **Evaluate**

| **a x b = 6** | | **a / b =** | |
|---|---|---|---|
| 1 x 6 | → | 1/6 = 0.167 | **A** |
| 2 x 3 | → | 2/3 = 0.667 | **B** |
| 3 x 2 | → | 3/2  =  1.5 | **C** |
| 6 x 1 | → | 6/1  =  6.0 | **E** |

**we don't see 1.667 here ( choice  D)**

**29. D**

Method: **Evaluate**

Evaluate $\dfrac{1-\dfrac{2}{x+1}}{1+\dfrac{3}{2x-5}}$  by putting **x = 0**

$$\frac{1-\dfrac{2}{0+1}}{1+\dfrac{3}{2(0)-5}} = \frac{5}{2}$$

Now if we put x=0 in the choice we should be -5/2

| (A) | $\dfrac{2x-5}{2x-1}$ | → | 5 |
|---|---|---|---|
| (B) | $\dfrac{2x-5}{2x+2}$ | → | -5/2 |
| (C) | $\dfrac{2x+5}{x-1}$ | → | -5 |
| (D) | $\dfrac{2x+5}{2x+2}$ | → | 5/2 |
| (E) | $\dfrac{2x+5}{2x+1}$ | → | 5 |

## 30. A

Method: **Evaluate**

$$a@b = \frac{ab^2}{a^b}$$

**(2 @ 4) - (3 @ 2 )**

$$\frac{2 \cdot 4^2}{2^4} - \frac{3 \cdot 2^2}{3^2} = \frac{2}{3}$$

## 31. C

Method: **Draw out and Evaluate**

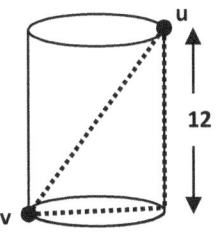

Apply Pythagoras

$$uv^2 = 6^2 + 12^2$$

**uv = 13.4**

## 32. D

Method: **Evaluate**

|b - a| = greatest

Let a = 1

|        | b  | \|b - a\| |
|--------|----|-----------|
| A      | 3  | 2         |
| B      | 5  | 4         |
| C      | 0  | 1         |
| D      | -5 | 6         |
| E      | -3 | 4         |

MAX occurs at **b = -5**

## 33. B

Method: **Draw out the triangle**

sin Z = $\frac{opp}{hyp}$ = $\boxed{\frac{XY}{ZX}}$

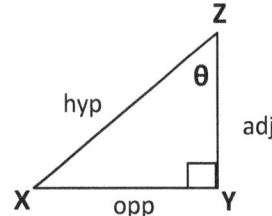

## 34. E

Method: **Evaluate**

In choice A and D matrices cannot be multiplied

In choice B and C matrices multiplication gives improper value of total sales of each brand

Choice E

$$|\;300 \quad 305 \quad 280\;| \quad \begin{vmatrix} 213 & 223 & 201 \\ 294 & 334 & 324 \\ 112 & 145 & 109 \end{vmatrix}$$

Day 1

300 x 213 + 305x294 + 280x112  =  184,930

Day 2

300 x 223 + 305x334 + 280x145  =  209,370

Day 3

300 x 201 + 305x324 + 280x109  =  189,640

|  **Day 1**  |  **Day 2**  |  **Day 3**  |
|-------------|-------------|-------------|
| \| 184,930  |  209,370    |  189,640 \| |

**35. C**

Method: **Evaluate**

| Old | Change | New |
|---|---|---|
| Mean = 80 | +3 → | Mean = 83 |
| Median = 76 | + 3 → | Median = 76 |
| S.D. = 6 | +0 → | S.D. = 6 |

**Changes occurs at the mean and median only**

**36. B**

Method: **Evaluate**

| | Distance | Time | Speed |
|---|---|---|---|
| **Autobahn** | 375 miles | 2.5 hours | 150 mph |
| **Town** | 120 miles | 1.5 hours | |
| **Total** | 495 miles | 4 hours | |

We work out the time by 4 - 2.5 = **1.5 hours**
and the distance by 495 - 375 = **120 miles**

Therefore **Speed** $= \dfrac{Distance}{Time} = \dfrac{120}{1.5} =$ **80 mph**

**37. C**

Method: **Draw out and Plug into the formula**

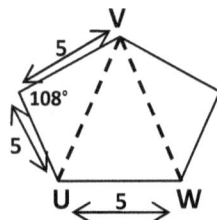

We need to find UV & VW

We know each angle is 540/5 = 108°

We know that UV = VW

Apply Cosine Rule

UV = VW $= \sqrt{5^2 + 5^2 - 2(5)(5)cos108} = 8.1$

We can find the height of triangle UVW by using Pythagoras:

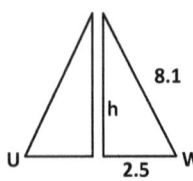

$2.5^2 + h^2 = 8.1^2$

h = 7.7

Area $= \dfrac{1}{2}$ x 2.5 x 7.7 = **9.63**

**38. D**

Method: **Evaluate**

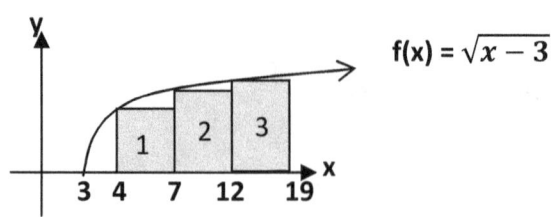

$f(x) = \sqrt{x-3}$

Area 1 = (7 - 4) $(\sqrt{4-3})$ = 3

Area 2 = (12 - 7) $(\sqrt{7-3})$ = 10

Area 3 = (19 - 12) $(\sqrt{12-3})$ = 21

**Total Area = 3 + 10 + 21 = 34**

**39. B**

Method: **Simplify**

$f(x) = \dfrac{2x+1}{1-x}$ → $y = \dfrac{2x+1}{1-x}$

swap x and y → $x = \dfrac{2y+1}{1-y}$

make y the subject → x(1 - y) = 2y + 1

x - xy = 2y + 1 → x - 1 = 2y + xy → x - 1 = y(2 + x)

$y = \dfrac{x-1}{2+x}$ → $f^{-1}(x) = \dfrac{x-1}{2+x}$

**40. D**

Method: **Expand**

middle term is ⟨?⟩ $x^2$ → $( 2x - 3 )^4$

$= \binom{4}{0} (2x)^4 + .... + \binom{4}{2} (2x)^2 (-3)^2 + ............$

$= ........................ + 6 \cdot (4x^2)(9) + ..............$

the middle term **is 216·$x^2$**

**41. B**

Method: **Solve using formula**

P(at least 2 tails)

= P(T T H)  or  P(T H T)  or  P(H T T)  or  P(T T T)

=  $(1/2)^3$  +  $(1/2)^3$ +  $(1/2)^3$ +  $(1/2)^3$

=  **1/2**

**42. D**

Method: **Simplify and Graph**

$5y - 3x > 0$  →  $5y > 3x$  →  **y > 0.6x**

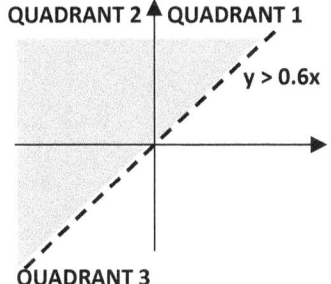

**Solutions lie in quadrant 1,2 and 3 only**

**43. E**

Method: **Plug into the formula and Graph**

We break the hexagon down into six equal triangles then we can find the area of each.

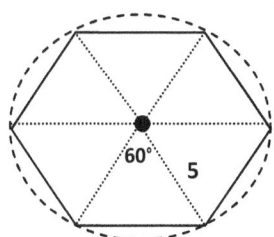

Area = $\frac{1}{2}$ a· b· sinC

Area = $\frac{1}{2}$ 5· 5· sin60  =  10.8

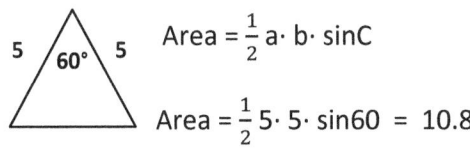

**Total area  =  6 x 10.8 = 64.9  or   65**

**44. E**

Method: **Graph**

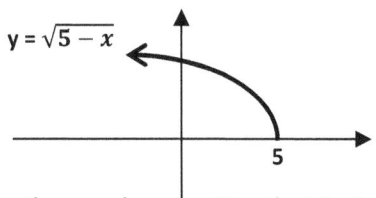

From the graph we notice that it starts at x =5 and move backward

**Domain :  x ≤ 5**

**45. A**

Method: **Simplify and Solve**

Let x be the length of the side of the cube

**surface area**  is **equal** to its **volume**

$6x^2$  =  $x^3$

x  =  6

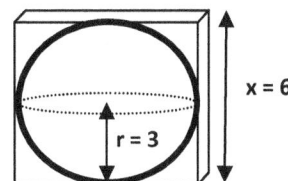

**V  = 4/3 π r$^3$ = 4/3 π 3$^3$  = 36π**

**46. B**

Method: **Evaluate using formula**

2, 1, 0.5, ......

We know that the sequence is geometric sequence with $U_1$ = **2** and **r = 0.5**

$S_{infinity}$  = $\frac{U_1}{1-r}$  = $\frac{2}{1-0.5}$ = **4**

## 47. D

Method: **Simplify**

$\ln(x + 3) = 2 \ln(3)$

$\ln(x+3) = \ln(3^2)$

$x + 3 = 3^2$

$x = 6$

## 48. D

Method: **Simplify and Plug into the identity**

$(\sin \theta - \cos \theta)^2 - 1 = \sin^2\theta - 2\sin\theta\cos\theta + \cos^2\theta - 1$

$= -2\sin\theta\cos\theta + \sin^2\theta + \cos^2\theta - 1 = -2\sin\theta\cos\theta + 1 - 1$

$= -2\sin\theta\cos\theta = \underline{-\sin 2\theta}$

## 49. B

Method: **Evaluate by drawing**

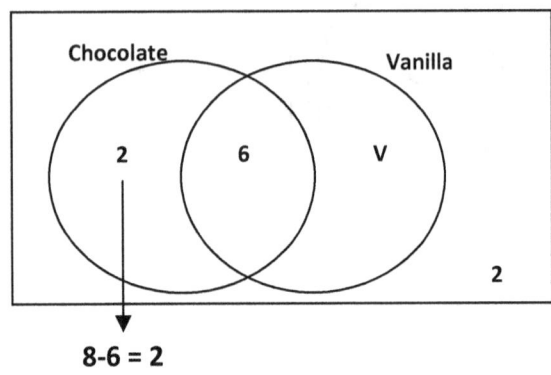

8-6 = 2

Total people = 20

$20 = 2 + 6 + V + 2$

V = **10 people** likes vanilla only

## 50. C

Method: **Solve using quadratic**

$y = x^2 + kx + 81$

We know that equal root means one roots, therefore the Discriminant must be equal to zero.

Discriminant → D = 0

$D = b^2 - 4ac$

$0 = k^2 - 4(1)(81)$

$k^2 = 324$

$k = \sqrt{324} = 18$

# Score Range

| Raw Score | Conversion |
|:---:|:---:|
| 45 - 50 | 800 |
| 40 - 44 | 750 - 790 |
| 36 - 39 | 720 - 740 |
| 30 - 35 | 690 - 710 |
| 25 - 29 | 650 - 680 |
| 20 - 24 | 590 - 640 |
| 16 - 19 | 540 - 580 |
| 12 - 15 | 510 - 530 |
| 9 - 11 | 450 - 500 |
| 5 - 10 | 400 - 440 |

Raw Score   =   Correct Answers  -  0.25 Wrong Answers

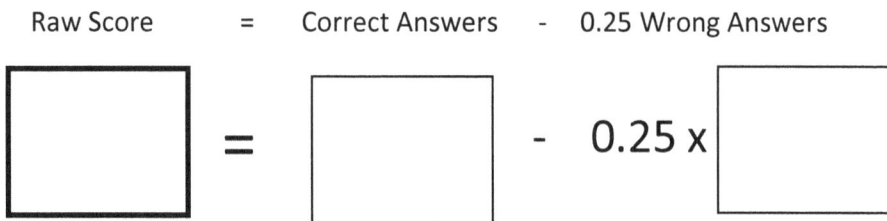

$$\boxed{\phantom{XX}} = \boxed{\phantom{XX}} - 0.25 \times \boxed{\phantom{XX}}$$

 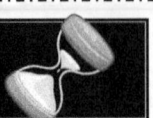
1. If $a^2 + b^2 = 25$ and $a^2 - b^2 = 7$ and then $2b^2 = ?$

(A)    16

(B)    18

(C)    26

(D)    32

(E)    44

2. The range of $f(x) = x^2 - 3$ is

(A)    Less than or equal to - 3

(B)    Greater than or equal to -3

(C)    Less than or equal to 3

(D)    Greater than or equal to 3

(E)    All real number

3. What is the first term of the sequence
        __ , **10, 30,60,100** ?

(A)    -10

(B)    -5

(C)    0

(D)    5

(E)    8

4. If $a^2 b^3 = 108$ , given that **a** and **x** are positive integer, then **ab** =

(A)    4

(B)    6

(C)    12

(D)    16

(E)    32

5. What is the x-intercept of the graph
        **5y - 2x = 10** ?

(A)    -5

(B)    -2

(C)    2

(D)    5

(E)    10

6. If $\dfrac{1}{\sqrt{x}+1} = 2$ , then x =

(A)    0

(B)    -1

(C)    1

(D)    4

(E)    no solution

7. If $\cos \theta = \sqrt{a}$ , then $\tan^2 \theta =$

(A)    $a^{0.5} - 1$

(B)    $a^{-1} - 1$

(C)    $a^{-2} + 1$

(D)    $a + 1$

(E)    $a - 1$

8. The sequence $\frac{2a}{5b}, \frac{a}{2b}, \frac{6a}{11b}, ....$ has the $n^{th}$ term equal to

(A)    $\frac{(n-1)a}{(5n-3)b}$

(B)    $\frac{(n-1)a}{(3n-2)b}$

(C)    $\frac{(n+1)a}{(3n-2)b}$

(D)    $\frac{(2n)a}{(3n+2)b}$

(E)    $\frac{(2n)a}{(3n-2)b}$

9. Matrix A has a dimension of 6 by 3 , matrix B has a dimension of 1 by 4 , matrix C has a dimension of 4 by 1 and matrix D has a dimension of 3 by 4   Which of the following is possible ?

(A)    ADB

(B)    ABC

(C)    BAC

(D)    BDA

(E)    DCB

10. The limit of $\lim_{n \to \infty} \left(1 - \frac{1}{n}\right)^2 =$

(A)    $-\infty$

(B)    $\infty$

(C)    -1

(D)    0

(E)    1

11. If $f(x) = 5^{x+1} - 5^{-x}$ , then $f^1(0) =$

(A)    - 2

(B)    - 1

(C)    -0.5

(D)    0

(E)    1

12. Which of the following point lies between $\frac{2}{9}$ and $\frac{2}{5}$ ?

(A)    0.12

(B)    0.21

(C)    0.39

(D)    0.52

(E)    0.63

SAT MATH LEVEL 1 Practice-Test

13. Every four years the population of a certain town increase by 30%, in year 2009 there were 400,000 people in the town. Which of the following expression must be used to indicate the population of the town in year 2017 ?

A.      $(400000)(8)(0.3)$

B.      $(400000)(1.3)^2$

C.      $(400000)(1.3)^8$

D.      $(400000)(1.3)^{17}$

(E)     $(400000)(8)(1.3)$

14. Which expression can be used to determine the height of the people using the amusement park ride ? Given that each person should be taller than 150 cm but lesser than 190 cm.

(A)     $|Height - 140| > 20$

(B)     $|Height - 140| < 20$

(C)     $|Height - 150| < 30$

(D)     $|Height - 160| > 20$

(E)     $|Height - 170| < 20$

15. If a triangle ABC is formed by the line **$y = 3x + 2$** , **$y = -2x + 12$** and the y-axis then what is the area of this triangle ?

(A)     20

(B)     12

(C)     10

(D)     8

(E)     6

16. What is the equation of axis of symmetry of the graph with equation **$y = x^2 - 6x + 12$** ?

(A)     $x = -6$

(B)     $x = 3$

(C)     $x = -12$

(D)     $y = 12$

(E)     $y = 3$

17. The blueprint of a village is drawn in a scale of 1.5 inches to 6 feet, a footpath is 18 inches long on a blueprint. What is the actual distance of the footpath ?

(A)     45      feet

(B)     63      feet

(C)     72      feet

(D)     81      feet

(E)     108     feet

18. Let **a** be a constant, the line perpendicular to line **$ay = 2ax + a^2$** is

(A)     $y = -ax + 2$

(B)     $y = -2x + a$

(C)     $y = -0.5x$

(D)     $y = \frac{-x}{a} + 2$

(E)     $y = \frac{x}{a} - a$

SAT MATH LEVEL 1 Practice-Test

19. Two square papers, with side of 5cm, are connected from the bases are glued and folded from the side to form a cylinder.

What is the volume of this cylinder ?

(A)      10

(B)      20

(C)      24

(D)      32

(E)      37

20. The graph of $y = x^2 - 3x$ intersects the graph of $y = 8 - 5x$ at point(s)

(A)      ( 2 , -2)  and     (4, 28)

(B)      ( 2 , -2)  and     (-4, 28)

(C)      ( -2 , 2)  and     (-4, -28)

(D)      ( -2 , -2)  and    (4, 28)

(E)      ( 2 , -2)  and     (-4, -28)

21.  If   $\sqrt[3]{-\dfrac{12}{3b^2}} = -1$   then b =

(A)      -4

(B)      $-\dfrac{1}{2}$

(C)      $-\dfrac{1}{4}$

(D)      $\dfrac{1}{4}$

(E)      2

22.  How many positive integers are in the solutions of  $|3x - 4| \leq 3$ ?

(A)      4

(B)      3

(C)      2

(D)      1

(E)      0

23.  If $\log_2 x = b$  and  $\log_2 y = c$  then  $2xy =$

(A)      $2^{b+c}$

(B)      $2^{bc}$

(C)      $2^{2+bc}$

(D)      $2^{2bc}$

(E)      $2^{b+c+1}$

24.  A clinic has a monthly budget of $ 54,000 to deal with all its expenses.  It needs to hire 'd' doctors at the rate of $ 7,000 per month and 'n' nurses at the rate of $3,000 per month to operate effectively.  Which of the following expression can the owner use to calculate the amount of doctors and nurses required ?

(A)      $7d + 3n \leq 54$

(B)      $7d + 3n \geq 54$

(C)      $3d \times 7n \leq 54$

(D)      $(7)(3)(d + n) \geq 54$

(E)      $(7)(3)(d + n) \leq 54$

SAT MATH LEVEL 1 Practice-Test

25. A coffee shop menu lists 5 type of coffees, 4 type of appetizers, and 3 type of spaghettis. Each day the coffee shop offers a combination of one of each at a special price. How many combination are possible when one of each kind is taken?

(A)     12

(B)     15

(C)     20

(D)     60

(E)     72

26. Which of the following is the equation of the graph below ?

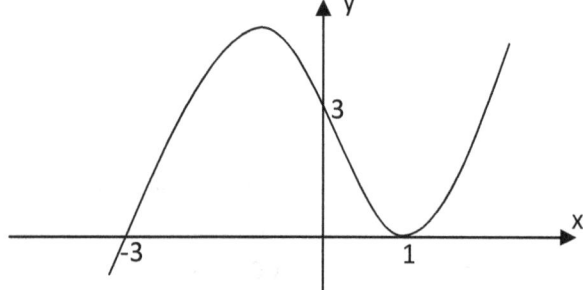

(A)     $x(x + 1)(x - 3)$

(B)     $x(x - 1)(x + 3)$

(C)     $(x - 1)^2 (x + 3)$

(D)     $(x + 1)^2 (x - 3)$

(E)     $x^2 (x - 1)(x + 3)$

27. The remainder when $ax^3 + 2x^2 - 18$ is divided by $(x - 3)$ is 9, then a =

(A)     1/3

(B)     1/2

(C)     1

(D)     2

(E)     3

28. If $\sec (x) = 1.25$ then $\tan (x) =$

(A)     0.5

(B)     0.75

(C)     0.87

(D)     1.33

(E)     1.57

29. The perimeter of an isosceles right triangle is equal to the perimeter of a square. If the square has an area of 9 cm$^2$, then what is the area of the triangle(to the nearest tenth) ?

(A)     4.5

(B)     6.1

(C)     7.3

(D)     8.4

(E)     10.2

30. Evaluate $(1 + i)^2(1 - i)^2$

(A)    -4

(B)    -2

(C)    0

(D)    2

(E)    4

31. If $\dfrac{5 \cdot n!}{(n-2)!} = 100$ , then **n =**

(A)    8

(B)    7

(C)    6

(D)    5

(E)    4

32. Given that

$$f(x) = \begin{cases} \mathbf{-2sinx} & -\pi \le x < 0 \\ \mathbf{2sinx} & 0 \le x < \pi \\ \mathbf{x - \pi} & x \ge \pi \end{cases}$$

What is the range of f(x) ?

(A)        $-\pi \le f(x) \le \pi$

(B)        $-2 \le f(x) \le 2$

(C)        $0 < f(x) < \pi$

(D)        $f(x) \ge 0$

(E)        $0 > f(x)$

33. Which of the following function represents the inverse function of the graph drawn below ?

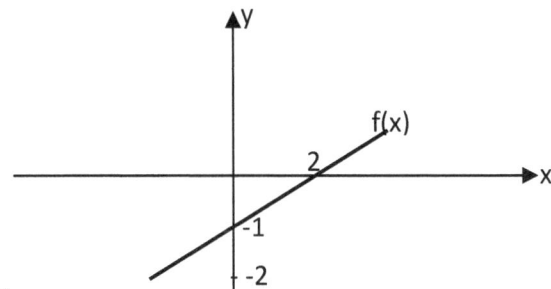

(A)    2x + 2

(B)    2x - 2

(C)    0.5 x - 1

(D)    0.5 x + 1

(E)    -0.5 x - 2

34. The period of graph **cos(a·x - b) + d** is

(A)    a·π

(B)    b·π

(C)    2π

(D)    2π/a

(E)    2π/b

35. $\dfrac{1-\dfrac{3x}{x-2}}{1+\dfrac{3}{x^2-4}} =$

(A) $\dfrac{2x+2}{2x-1}$

(B) $\dfrac{2x-2}{2x+2}$

(C) $\dfrac{2x+4}{2x-1}$

(D) $\dfrac{-2x+2}{x+1}$

(E) $\dfrac{-2x-4}{x-1}$

36. The fifth term of arithmetic sequence is -9 and the tenth term is -19, the first term is equal to

(A)    -1

(B)    1

(C)    3

(D)    7

(E)    9

37. Which of the following is the graph of a function ?

(A)    x = 5

(B)    $x^2 + y^2 = 16$

(C)    8x = y

(D)    $y^2 = 6x$

(E)    $4x^2 + 5y^2 - 100 = 0$

38. If $f(x) = \ln(x+5)$ then $f^{-1}(0) =$

(A)    -5

(B)    -4

(C)    0

(D)    5

(E)    ln( 5)

39. Which of the following statement must be true ?

I. The mean of the set is always greater than standard deviation

II. The variance is always greater than standard deviation

III. Mode is always greater than the mean

(A)    I only

(B)    II only

(C)    III only

(D)    I and II only

(E)    II and III only

40. The probability that Lee passes a math exam is 0.2 and the probability that Lee fails a chemistry exam is 0.3. What is the probability that Lee pass at least one of the subject ?

(A)    0.86

(B)    0.76

(C)    0.61

(D)    0.14

(E)    0.06

41. If Josh invested \$4,000 at the rate of 5% compound interest then how long (to the nearest year) would it take for the money to double its value ?

(A)     8

(B)     10

(C)     12

(D)     14

(E)     16

42. If $\sqrt{a} \times b = 10$ and $a \times \sqrt{b} = 20$ then $\dfrac{a}{b} =$

(A)     0.25

(B)     0.5

(C)     2

(D)     4

(E)     16

43. The average of three numbers is 12 and one of the number is twice the other. If the range is 5 then what is the value of the largest number ?

(A)     3.5

(B)     6.75

(C)     7.5

(D)     9.75

(E)     12.75

44. Given that **f(a) = 28** and the area bounded by the graph of **f(x)** from $0 \le x \le a$ is 32 then f(x) is equal to

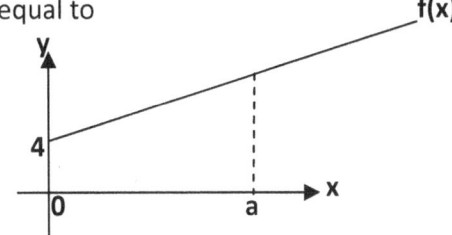

(A)     12x + 4

(B)     4x + 4

(C)     2x + 4

(D)     x + 4

(E)     0.25x + 4

45. A right triangle with the area of 24 has longest side equal to 9. What is the perimeter of this triangle

(A)     $9 + \sqrt{17} + \sqrt{19}$

(B)     $9 + \sqrt{177}$

(C)     $12 + \sqrt{17}$

(D)     $17 + \sqrt{177}$

(E)     $\sqrt{17} + \sqrt{177}$

46. What is the domain of **f(x) = $ln(x - 2)$** ?

(A)     $0 < x \le 2$

(B)     $2 < x < \infty$

(C)     $-\infty > x > 2$

(D)     $-2 \ge x \ge 2$

(E)     $-2 \le x \le 2$

47. sin (x - π) =

(A)    cos x

(B)    sin x

(C)    - cos x

(D)    - sinx

(E)    tan x

48. If $g(x) = \frac{2e^x - 10}{e^{2x} - 3e^x}$ and $f(x) = \frac{2(x-5)}{x(x-1)}$ then
    f(g(0)) =

(A)    $\frac{-1}{6}$

(B)    -6

(C)    $e^6$

(D)    6

(E)    $\frac{1}{6}$

49. If $x - 9 = 2(\sqrt{x} + 3)$ , then x =

(A)    1 only

(B)    9 only

(C)    25 only

(D)    25 and 9

(E)    no solution

50. Which of the following is the equation of the function shown in the table below ?

| x    | 0 | 1  | 2  | 3 | 4 | 5  |
|------|---|----|----|---|---|----|
| f(x) | 0 | -2 | -2 | 0 | 4 | 10 |

(A)    f(x) = x - 3

(B)    f(x) = $x^2$ - 2x

(C)    f(x) = x - $2x^2$

(D)    f(x) = $x^2$ + 2x

(E)    f(x) = $x^2$ - 3x

# END OF TEST 3

## 1. B

Method: **Solve**

**Square both side**

$a^2 + b^2 = 25$

$a^2 - b^2 = 7$ minus

$2b^2 = 18$

## 2. B

Method: **Graph**

$f(x) = x^2 - 3$

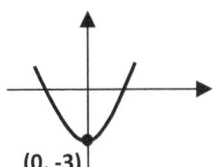

(0, -3)

Range: $y \geq -3$

## 3. C

Method: **Find the pattern**

| Term | 1 | 2 | 3 | 4 | 5 |
|---|---|---|---|---|---|
| Sequence | X | 10 | 30 | 60 | 100 |
| Increment | +10 | +20 | +30 | +40 | |

We found that the sequence is increased by 10 , 20 , 30 , 40 and so on. So we let the first term be X then X + 10 = 10 → **X = 0**

## 4. B

Method: **Solve by factorizing**

$a^2 b^3 = 108$

$a^2 b^3 = 4 \times 27$

$a^2 b^3 = 2^2 \times 3^3$

Now we know that a = 2 and b = 3

**a·b = 2 x 3 = 6**

## 5. A

Method: **Evaluate**

**x-intercept → y = 0, x = ?**

$5y - 2x = 10$

$5(0) - 2x = 10$

x = -5

## 6. E

Method: **Solve**

$\dfrac{1}{\sqrt{x}+1} = 2 \rightarrow 1 = 2\sqrt{x} + 2$

$-1 = 2\sqrt{x}$

$\dfrac{-1}{2} = \sqrt{x}$

$x = \dfrac{1}{4}$

SAT MATH LEVEL 1 Practice-Test

## 7. B

Method: **Simplify using trig**

$$\cos \theta = \frac{adj}{hyp} = \frac{\sqrt{a}}{1}$$

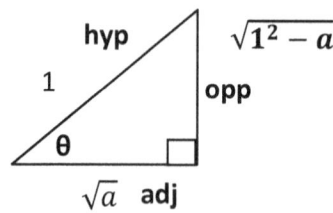

$$\tan \theta = \frac{opp}{adj} = \frac{\sqrt{1-a}}{\sqrt{a}}$$

$$\tan^2 \theta = \frac{1-a}{a} = a^{-1} - 1$$

## 8. D

Method: **Use formula**

Sequence $\frac{2a}{5b}, \frac{a}{2b}, \frac{6a}{11b}, ....$

same as $\frac{2a}{5b}, \frac{4a}{8b}, \frac{6a}{11b}, ....$

- - - - - - - - - - - - - - - - - - - - - - - - -
numerator sequence → 2a, 4a, 6a, 8a,....

n$^{th}$ term → (2n)a
- - - - - - - - - - - - - - - - - - - - - - - - -
denominator sequence → 5b, 8b, 11b, 14b, ....

n$^{th}$ term → (3n+2)b
- - - - - - - - - - - - - - - - - - - - - - - - -

combining the two → n$^{th}$ term $= \dfrac{(2n)a}{(3n+2)b}$

## 9. E

Method: **Evaluate**

D x C is possible because

Column of D  = Row of C

D x C  has a dimension of 3 by 1

DC x B is possible because

Column of DC  =  Row of B

DCB has a dimension of 3 by 4

## 10. E

Method: **Evaluate and Simplify**

$$\lim_{n\to\infty} \left(1 - \frac{1}{n}\right)^2 = \left(1 - \frac{1}{\infty}\right)^2 = (1-0)^2 = \underline{\underline{1}}$$

## 11. C

Method: **Evaluate**

$f(x) = 5^{x+1} - 5^{-x}$  →   $y = 5^{x+1} - 5^{-x}$

$f^1(0)$  = ??  →   if **y = 0** then **x = ??**

$$y = 5^{x+1} - 5^{-x}$$

$$0 = 5^{x+1} - 5^{-x}$$

$5^{-x} = 5^{x+1}$  ← same base so power are equal

$-x = x + 1$ → $-2x = 1$ → $x = \underline{\textbf{-0.5}}$

## 12. **C**

Method: **Graph**

Numbers between $\frac{2}{9}$ or 0.22222... and $\frac{2}{5}$ or 0.4

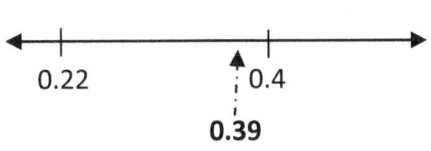

## 13. **C**

Method: **Evaluate**

year 2009 to 2017 means **t = 8**

this is exponential growth function with initial value of 400,000 and growth rate of 1.3

Total population = initial value x (rate)$^{year}$

**Total population = (400,000)(1.3)$^7$**

## 14. **E**

Method: **Evaluate**

We want the height to be between
        150 < height < 190

difference between 190 and 150 is **20**

        -20 < height - M < 20

M is the middle number between 150 and 190 which is 170   **(M = 170)**

-20 < height - 170 < 20 → **|height - 170| < 20**

## 15. **C**

Method: **Graph out and Evaluate**

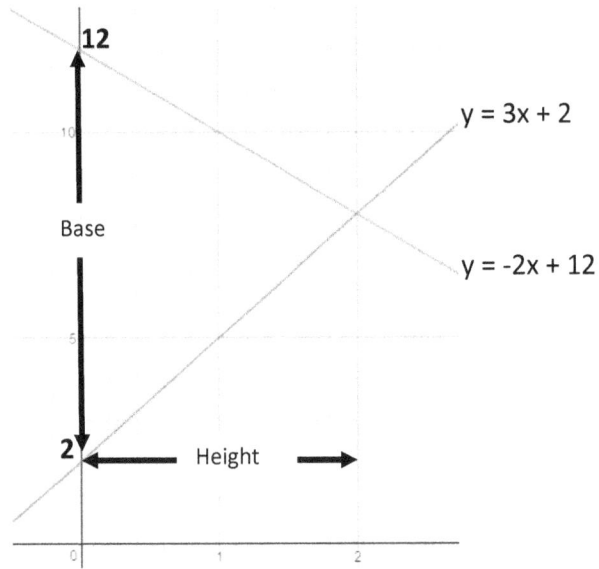

Base  = 12 - 2 = 10

Height =  2 - 0 = 2

**Area = 0.5 x 10 x 2  = 5**

## 16. **B**

Method: **Graph**

y = x$^2$ - 6x + 12  → complete the square for vertex

y = (x - 3)$^2$ + 3   **→ vertex at (3, 3)**

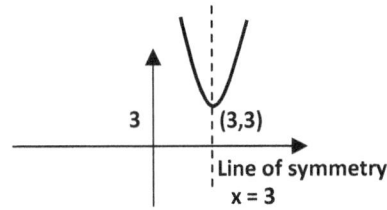

**17. C**

Method: **Evaluate using ratio**

$$\frac{1.5\ inch}{6\ ft} = \frac{18\ inch}{x\ ft}$$

$$x = \frac{18 \times 6}{1.5} = \textbf{72 feet}$$

**18. C**

Method: **Simplify**

$ay = 2ax + a^2 \rightarrow y = 2x + a$

Slope = 2

Slope of perpendicular = -1/2 or = -0.5

The only equation with this slope is

**y = - 0.5x**

**19. B**

Method: **Draw out and Evaluate**

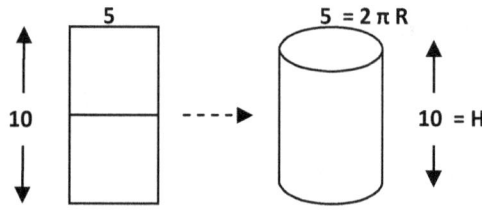

5 = circumference

$5 = 2\pi R \rightarrow$ **R = 0.8**

Volume $= \pi R^2 H = \pi (0.8)^2 (10)$

**Volume = 20 cm$^3$**

**20. B**

Method: **Solve**

$y = x^2 - 3x$ and $y = 8 - 5x$ are equals to each other if they intersect

$$x^2 - 3x = 8 - 5x$$

$$x^2 + 2x - 8 = 0$$

$$(x - 2)(x + 4) = 0$$

| x = 2 | and | x = -4 |
|-------|-----|--------|
| y = -2 | | y = 28 |

**So they meet at (2, -2) and (-4, 28)**

**21. E**

Method: **Solve**

$\sqrt[3]{-\dfrac{12}{3b^2}} = -1$  ← simplify and cube both side

$\sqrt[3]{-\dfrac{4}{b^2}} = -1 \rightarrow b^2 = 4 \rightarrow$ **b = ± 2**

**22. C**

Method: **Solve**

integers are in the solutions of **|3x - 4| ≤ 3**

$$-3 \leq 3x - 4 \leq 3$$

$$1 \leq 3x \leq 7$$

$$0.33 \leq x \leq 2.33$$

**integer value of x = { 1 , 2 }**

**only 2 values**

**23. E**

Method: **Plug into the formula**

$\log_2 x = b \rightarrow 2^b = x$

$\log_2 y = c \rightarrow 2^c = y$

$2xy = 2(2^b)(2^c) = 2^{1+b+c}$

$= 2^{b+c+1}$

**24. A**

Method: **Evaluate**

maximum budget is $ 54,000

$\qquad$ total cost $\leq$ 54,000

doctors wage + nurses wage $\leq$ 54,000

$\qquad$ 7000(d) + 3000 (n) $\leq$ 54,000

$\qquad$ **7d + 3n $\leq$ 54**

**25. D**

Method: **Simplify using formula**

this question is about combination

5 coffee x 4 appetizers x 3 spaghettis

= **60 combinations**

**26. C**

Method: **Simplify using formula**

the roots are 1 and -3 put them in the formula

$y = (x - 1)^2(x + 3)$ $\leftarrow$ the y-intercept is +3

**27. A**

Method: **Evaluate**

Using remainder theorem

$R(x) = ax^3 + 2x^2 - 18 \rightarrow R(3) = 9$

$\quad 9 = a(3)^3 + 2(3)^2 - 18$

$27a = 9$

$\quad a = \underline{1/3}$

**28. B**

Method: **Plug into the formula**

$\qquad\qquad \sec(x) = 1.25$

$\qquad\qquad\qquad\qquad\qquad$ fraction

$\sec(x) = \dfrac{hyp}{adj} \rightarrow \sec(x) = \dfrac{5}{4}$

Draw out triangle

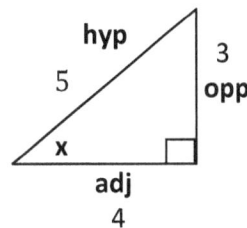

$\tan(x) = \dfrac{opp}{adj} = \dfrac{3}{4} = 0.75$

**29. B**

Method: **Evaluate by drawing**

Area of square → $x^2 = 9$

Side of square → $x = 3$

Perimeter of square = $4x = 4(3) = $ **12**

Perimeter of right isosceles triangle = **12**   (also)

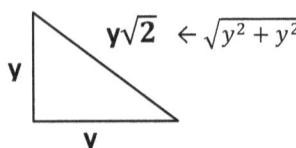

$y\sqrt{2} \leftarrow \sqrt{y^2 + y^2}$

Perimeter of triangle = $y + y + y\sqrt{2}$

$$12 = 2y + y\sqrt{2}$$

$$y = 3.5$$

Area $= \frac{1}{2} \times Y \times Y = \frac{1}{2} \times 3.5 \times 3.5 = $ **6.1 cm$^2$**

**30. E**

Method: **Simplify by expansion**

$(1 + i)^2(1 - i)^2 = (1^2 + 2(1)(i) + i^2)(1^2 - 2(1)(i) + i^2)$

$= (1 + 2i - 1)(1 - 2i - 1) = (2i)(-2i)$

$= (-4i^2) = \underline{\underline{4}}$

**31. D**

Method: **Evaluate**

$$\frac{5n!}{(n-2)!} = 100 \rightarrow \frac{n(n-1)(n\cancel{-2})!}{(n\cancel{-2})!} = 20$$

$n(n-1) = 20 \rightarrow n^2 - n - 20 = 0$

$$(n-5)(n+4) = 0$$

$$\boxed{n = 5} \text{ and } n = -4$$

**32. D**

Method: **Graph**

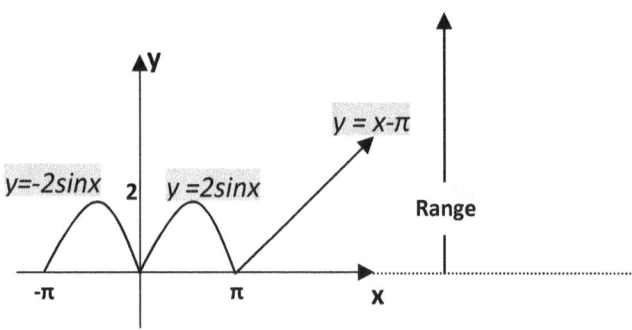

Range : **f(x) ≥ 0**

**33. A**

Method: **Plug into the formula**

We know that f(x) is a linear equation with a slope of $\frac{1}{2}$ and y-intercept of -1

$$f(x) = \frac{1}{2}x - 1$$

Find inverse   **Step1:**      $y = \frac{1}{2}x - 1$

**Step 2** (swap x and y ):      $x = \frac{1}{2}y - 1$

**Step 3** (make y the subject):   $x = \frac{1}{2}y - 1$

$$y = 2(x + 1)$$

Therefore, $f^{-1}(x) = 2x + 2$

**34. D**

Method: **Evaluate**

$\cos(a \cdot x - b) + d$ → period of cot(x) is same as tan(x)

**Period** $= \frac{2\pi}{a}$

## 35. E

Method: **Plug in value of x**

**use short-cut by letting x = 0**

$$\dfrac{1-\dfrac{3x}{x-2}}{1+\dfrac{3}{x^2-4}} \quad \rightarrow \quad \text{put } x = 0 \rightarrow 4$$

(A) $\dfrac{2x+2}{2x-1}$ → put x = 0 → -2

(B) $\dfrac{2x-2}{2x+2}$ → put x = 0 → -1

(C) $\dfrac{2x+4}{2x-1}$ → put x = 0 → -4

(D) $\dfrac{-2x+2}{x+1}$ → put x = 0 → 2

(E) $\dfrac{-2x-4}{x-1}$ → put x = 0 → 4

## 36. A

Method: **Evaluate using formula**

arithmetic sequence formula

**$U_n = U_1 + (n-1)d$**

$U_5 = U_1 + (5-1)\,d \rightarrow$ **-9 = $U_1$ + 4d**

minus

$U_{10} = U_1 + (10-1)\,d \rightarrow \underline{\textbf{-19} = \textbf{U}_1 + \textbf{9d}}$

$$10 = -5d \rightarrow d = -2$$

Put d = -2 in $U_5$ equation to find $U_1$

$-9 = U_1 + 4d \rightarrow -9 = U_1 + 4(-2)$

**$U_1 = -1$**

## 37. C

Method: **Evaluate using vertical line test**

A.  x = 5 ← many to one ( not a function)

B.  $x^2 + y^2 = 16$ ← many to many(not a function)

**C.  8x = y ← one to one ( a function)**

D.  $y^2 = 6x$ ← many to one(not a function)

E.  $4x^2 + 5y^2 = 16$ ← many to many(not a function)

## 38. B

Method: **Evaluate**

find $f^1(0)$ = ?? → y = 0 , x = ??

$$f(x) = \ln(x+5)$$

$$0 = \ln(x+5)$$

take 'e' both side → $e^0 = x + 5$

$$1 = x + 5$$

$$x = -4$$

## 39. A

Method: **Evaluate**

I.  The mean of the set is always greater than standard deviation → **This is always TRUE**

II.  The variance is always greater than standard deviation → **var = s.d.$^2$  (not always true)**

III.  Mode is always greater than the mean
→**Not necessary always true**

## 40. B

Method: **Evaluate using formula**

Probability that Lee passes a math exam is 0.2

Probability that Lee fails a math exam is 0.8

Probability that Lee fails a chem. exam is 0.3

Probability that Lee passes chem. exam is 0.7

P(Lee pass at least one exam )

$\quad$ = 1 - P(fail all)

$\quad$ = 1 - P(fail math) x P(fail chem)

$\quad$ = 1 - 0.8 x 0.3

$\quad$ = 1 - 0.24 = **0.76**

## 41. D

Method: **Evaluate using formula**

Total money = principal $\cdot(1 + \text{rate})^{\text{time}}$

$\quad$ 8000 $\quad$ = $4000 ( 1 + 0.05)^{\text{time}}$

$\quad\quad$ 2 $\quad$ = $1.05^{\text{time}}$

$\quad$ ln (2) $\quad$ = $\ln ( 1.05^{\text{time}})$

$\quad\quad$ time = 14.2 years

It will take about **14 years** for the money to double its values.

## 42. D

Method: **Simplify**

$\frac{a}{b}$ = ??

square both side

$\frac{\sqrt{a} \times b}{a \times \sqrt{b}} = \frac{10}{20} \rightarrow \frac{\sqrt{b}}{\sqrt{a}} = \frac{1}{2} \rightarrow \frac{b}{a} = \frac{1}{4}$

$\quad\quad \frac{a}{b}$ = **4**

## 43. E

Method: **Solve by listing**

Let $x < y < z$

$y = 2x$ , $\quad\quad$ z - x = 5 $\rightarrow$ z = 5+ x $\quad$ ,z = ??

$\frac{x + y + z}{3} = 12 \rightarrow$ Put y=2x in $\rightarrow$ x + 2x + z = 36

3x + z = 12 $\rightarrow$ Put z = 5+ x  in $\rightarrow$ 3x + 5+ x = 36

4x = 36 - 5 $\rightarrow$ x = 7.75

z = 5 + x $\rightarrow$ **z** = 5 + 7.75 = **12.75**

## 44. A

Method: **Evaluate the area**

Area of trapezium = $\frac{1}{2}$ x base ( $h_1 + h_2$)

Area = 0.5 x a x (4 + f(a)) = 0.5a(4 + 28)

32 = 0.5a(32) $\rightarrow$ a = 2

y-intercept = 4, slope = $\frac{rise}{run} = \frac{28-4}{2-0}$ = 12

Equation of **f(x) = 12x + 4**

**45. B**

Method: **Solve**

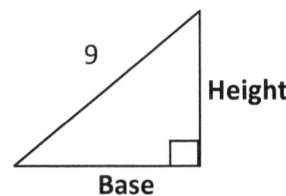

Area = 24 → **0.5 x base x height = 24**

So, Base x Height = 48 and Base$^2$ + Height$^2$ = 9$^2$

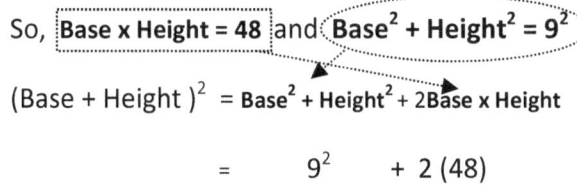

(Base + Height )$^2$ = Base$^2$ + Height$^2$ + 2Base x Height

= 9$^2$ + 2 (48)

(Base + Height )$^2$ = 177

Base + Height = $\sqrt{177}$

**Perimeter = 9 + Base + Height = 9 + $\sqrt{177}$**

**46. B**

Method: **Graph**

**f(x) = $l\,n(x-2)$**

x = 2

3

**Domain: 2 < x < ∞**

**47. D**

Method: **Simplify using formula**

sin (A - B) = sinAcosB - cosAsinB

sin (x - π) = sinx·cos(π) - cox·sin(π)

sin (x - π) = sinx (-1) - cosx (0) = **- sinx**

**48. A**

Method: **Evaluate**

**Find g(0) first**

g(0) = $\frac{2e^0-10}{e^0-3e^0}$ = $\frac{-8}{-2}$ = **4**

f(g(0)) = f(4) = $\frac{2(4-5)}{4(4-1)}$ = $\frac{-2}{12}$ = $\frac{-1}{6}$

**49. C**

Method: **Solve**

$x - 9 = 2(\sqrt{x} + 3)$ → x - 2$\sqrt{x}$ - 15 = 0

Factor → ($\sqrt{x}$ - 5 ) ($\sqrt{x}$ + 3 ) = 0

$\sqrt{x}$ = 5 → x = 25

$\sqrt{x}$ = -3 → x = 9

**If we put x = 25 back we get the real solution**

**50. E**

Method: **Evaluate each**

| x | 0 | 1 | 2 | 3 | 4 | 5 |
|---|---|---|---|---|---|---|
| f(x) | 0 | -2 | -2 | 0 | 4 | 10 |

Try putting in f(5) , the choice that gives 10 is the answer

| A. | f(5) = 2 | WRONG |
|---|---|---|
| B. | f(5) = 15 | WRONG |
| C. | f(5) = -45 | WRONG |
| D. | f(5) = 35 | WRONG |
| **E.** | **f(5) = 10** | **CORRECT** |

# Score Range

| Raw Score | Conversion |
|:---:|:---:|
| 47 - 50 | 800 |
| 42 - 46 | 750 - 790 |
| 38 - 41 | 720 - 740 |
| 34 - 37 | 690 - 710 |
| 30 - 33 | 650 - 680 |
| 26 - 29 | 590 - 640 |
| 23 - 27 | 540 - 580 |
| 18 - 22 | 510 - 530 |
| 13 - 17 | 450 - 500 |
| 7 - 11 | 400 - 440 |

Raw Score    =    Correct Answers    -    0.25 Wrong Answers

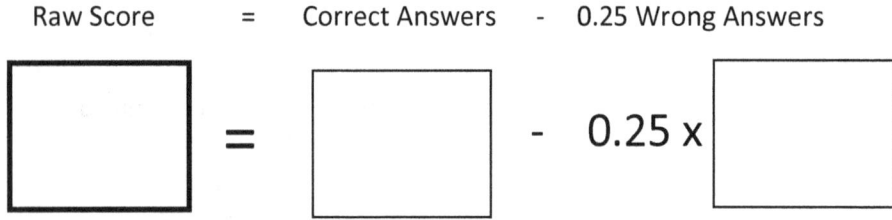

$$\boxed{\phantom{XX}} = \boxed{\phantom{XX}} - 0.25 \text{ x } \boxed{\phantom{XX}}$$

1. What is the slope from the origin to point A (9,12) ?

(A)     -3/4

(B)     -4/3

(C)     -5/6

(D)     4/3

(E)     3/4

2. If $(x + 2)^2 = (x - 2)^2$ then what value of x will make the statement true ?

(A)     -2 only

(B)     2 and -2 only

(C)     -4 only

(D)     0 only

(E)     no real result

3. Which of the following equation has x-intercept of 5 and slope of -1 ?

(A)     $y - x + 5 = 0$

(B)     $y + x - 5 = 0$

(C)     $5x + 5y = 1$

(D)     $5x - 5y = 1$

(E)     $5x = 5y$

4. A parallelogram with one angle equal to 40° has a perpendicular height equal to half of its base.     If the area is equal to 72 cm$^2$ then what is the perimeter of this parallelogram ?

(A)     72

(B)     42.66

(C)     21.33

(D)     18.44

(E)     9.33

5. If $\dfrac{2\sqrt{a+b}}{3} = 1$ , then $(a + b)^2 =$

(A)     $\dfrac{3}{2}$

(B)     $\dfrac{9}{4}$

(C)     $\dfrac{81}{16}$

(D)     $\dfrac{\sqrt{3}}{\sqrt{2}}$

(E)     $\dfrac{2}{3}$

6. If $x^2 > |x|$ , then

(A)     $x > 1$

(B)     $x < 1$

(C)     $0 < x < 1$

(D)     $-1 < x < 1$

(E)     $-1 > x > 1$

7. If $g(x) = e^x + 1$ then $g^{-1}(x) =$

(A)     ln (x - 1)

(B)     ln (x + 1)

(C)     ln (e - x)

(D)     ln (e + x)

(E)     ln ($e^x$ - 1)

8. If $ax^2 + bx - 5$ has root equal to **5** and **-1** then the value of **a + b = ?**

(A)     -5

(B)     -4

(C)     -3

(D)     3

(E)     4

9. The graph of $f(x) = x^2 + 2x + 1$ is translated vertically +4 units and horizontally -3 units, the vertex of will now be at

(A)     (4, -4)

(B)     (-4, 4)

(C)     (-4, -4)

(D)     (4 , 4)

(E)     ( 0, 4)

10. The limit of $\lim_{x \to 0}(e^{-x} + 3)^2 =$

(A)     -3

(B)     0

(C)     4

(D)     9

(E)     16

11. How many ways can 6 people can be lineup given that the tallest should stand in front and the shortest should stand the end ?

(A)     24

(B)     60

(C)     720

(D)     1680

(E)     40320

12. The graph has an equation of **y = a·sin(bx +c)**. Which of the following is the value of **b** ?

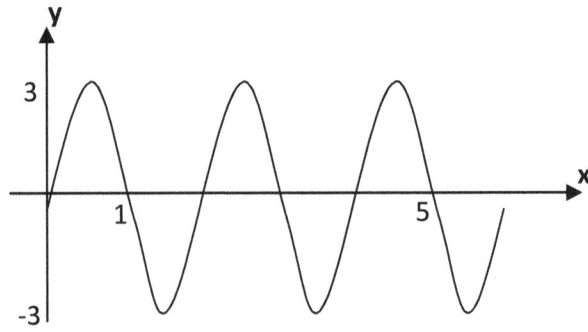

(A)     1/2          (B)     2

(C)     π/2          (D)     π

(E)     5

13. The graph of rational function

$y = \dfrac{8}{x^2-16}$ has vertical asymptote of x =

(A)     -16

(B)     - 4

(C)      2

(D)     -2 and 4

(E)     -4  and 4

14. Triangle ABC has the side of length 5, 12 and 13. To the nearest degree what is the measure of the smallest angle of triangle ABC ?

(A)     20

(B)     23

(C)     27

(D)     33

(E)     40

15.  A toy rocket fired up into the air is modeled by a function $h(t) = 5t - t^2$ , where h represents the height of the rocket and t represents the time in second.  At what time will the rocket reach the maximum height ?

 (A)     5 seconds

(B)     3 seconds

(C)     2.5 seconds

(D)     2  seconds

(E)     1.5 seconds

16.  A circle with equation $(x - 3)^2 + (y^2 +7) = 16$, which of the following is the value of the radius of this circle ?

(A)     2

(B)     3

(C)     4

(D)     9

(E)     16

17.  If u and v are the domain of a function g and g(u) < g(v), which of the following must be true ?

(A)     u = 0

(B)     u < v

(C)     u > v

(D)     u = v

(E)     u ≠ v

18.  If two circles of radius 3 cm and 5 cm are fitted inside a rectangle.  What is the area of the rectangle?

(A)     64

(B)     100

(C)     160

(D)     256

(E)     312

19. The line tangent to the vertex of
$y = (x-2)^2 + 3$ has the equation of

(A)    $y = -2$

(B)    $y = 3$

(C)    $y = -3$

(D)    $x = -2$

(E)    $x = 2$

20. If $\sqrt[5]{-\dfrac{32}{x^{15}}} = -16$ then x =

(A)    $-\dfrac{1}{2}$

(B)    $-\dfrac{1}{4}$

(C)    $\dfrac{1}{2}$

(D)    $\dfrac{1}{4}$

(E)    4

21. A parallelogram has the coordinates of
( 3,2) , (4,6) , (7,6) and (a,b). Which of the
following represents the coordinate and the
area of the parallelogram correctly ?

|    | Coordinate | Area |
|----|-----------|------|
| A. | (2,6)     | 12   |
| B. | (6,2)     | 6    |
| C. | (2,6)     | 6    |
| D. | (6,2)     | 12   |
| E. | (2,-6)    | 18   |

22. What is the sum of the sequence
1,2,2,3,3,3,4,4,4,4,..........10 ?

(A)    302

(B)    324

(C)    362

(D)    385

(E)    390

23. What is the equation of the line that is
perpendicular to line 3x + 2y = 12 ?

(A)    2x - 3y = 9

(B)    3x - 2y = -3

(C)    -2x - 3y = 3

(D)    3x = -12

(E)    -2y = 4

24. The solution of $2x^2 - 7x + 3 = 0$ is

(A)    x = -0.5  and  x = 3

(B)    x = 0    and  x = -3

(C)    x = 0    and  x = -0.5

(D)    x = 0.5  and  x = 3

(E)    x = -0.5  and  x = -3

SAT MATH LEVEL 1 Practice-Test

25. The range of graph **2cos(x) - 9** is

(A)     [-9 , -7]

(B)     [-7,  7]

(C)      [-9,  -7]

(D)     [-7,  9]

(E)      [-11,  -7]

26. The average of three numbers is 50 and one of the number is twice the other and the last number is half the other.  What is the value of the smallest number ?

(A)     10

(B)     20

(C)     25

(D)     30

(E)     60

27.  The graph of **y = -x$^2$ + 4** and **y = -x$^3$** intersect each other

(A)     0     time

(B)     1     time

(C)     2     times

(D)     3     times

(E)     4     times

28.  The remainder when  **4x$^2$ + 2x - 1** is divided by **(x + 1)** is

(A)     -2

(B)     -1

(C)     0

(D)     1

(E)     5

29.  The eight term of arithmetic sequence is $\sqrt{5}$ and the tenth term is $3\sqrt{5}$ , the first term is equal to

(A)     $-6\sqrt{5}$

(B)     $-3\sqrt{5}$

(C)     $-\sqrt{5}$

(D)     $\sqrt{5}$

(E)     5

30. For **0 <  θ < π** the value of
**0.5cos$^2$(θ) + 0.5sin$^2$(θ) =**

(A)     0

(B)     0.5

(C)     1

(D)     2

(E)     π

SAT MATH LEVEL 1 Practice-Test

31. The value of $(8 - 6i)^2$ is equal to

(A)      28 + 96i

(B)      28 - 96i

(C)      100 - 96i

(D)      100 - 28i

(E)      100 + 28i

32. What is the value of the coefficient of $x^3$ in the expansion of $(2 + 3x)^4$ ?

(A)      216

(B)      128

(C)       72

(D)       64

(E)       27

33. If $(n - 2)! = (n - 3)! \cdot 4!$ , then n =

(A)      26

(B)      24

(C)      23

(D)      21

(E)      18

34. A car depreciates its value by 10% each year after the first year, given that the first year it's price goes down by 30%. If Jill bought a BMW M5 for fifty-thousand dollar and drive it for 4 years, how much would it worth ?

(A)      32805

(B)      31280

(C)      28790

(D)      25515

(E)      22964

35. cos(arctan(0.25)) =

(A)      4.12

(B)      1.03

(C)      0.97

(D)      0.71

(E)      0.25

36. If $\log_a 312 = 7$, then a =

(A)      4.22

(B)      2.27

(C)      1.76

(D)      0.79

(E)      0.52

SAT MATH LEVEL 1 Practice-Test

37. Given that

$$f(x) = \begin{cases} e^x & , x < 1 \\ \ln(x) & , 1 \le x < 10 \\ \ln(e^{-x+10}) & , x \ge 10 \end{cases}$$

What is the range of f(x) ?

(A)        $0 \le f(x) < \infty$

(B)        $-\infty < f(x) \le 0$

(C)        $-\infty < f(x) < \infty$

(D)        $f(x) > \ln(10)$

(E)        $\ln(10) > f(x)$

38. Which of the following statement must be true about the data set {1,2,2,3,3,3,4,4,4,4}?

**I. The range of the data is 4**

**II. The variance is less than the mean**

**III. The mean is equal to 3**

(A)     I only

(B)     II only

(C)     III only

(D)     I and II only

(E)     II and III only

39. An artist can paint a circular wall with radius of 1.5 meters in 2 hours, how long would it take an artist to paint square wall with length of 3 meters ?

(A)     1.5

(B)     2.5

(C)     3.5

(D)     4.0

(E)     5.0

40. The equation of the circle

$(y-4)^2 + (x-3)^2 = 36$ has the center and radius of

|       | Center    | Radius |
|-------|-----------|--------|
| (A)   | (4,3)     | 6      |
| (B)   | (3,4)     | 6      |
| (C)   | (-4, -3)  | 18     |
| (D)   | (-3,-4)   | 18     |
| (E)   | (3,4)     | 36     |

41. If $x_0 = 0.5$ and $x_{n+1} = (x_n)^2 (x_n + 2)$ then to the nearest integer $x_3 =$

(A)     52

(B)     46

(C)     12

(D)     3

(E)     1

**42.** The probability of picking a red candy from the box is one-fifth and the probability of picking a blue candy is two-third. If there were red, blue and green candies in the box then which of the following could be the total number of candies in the box ?

(A)     9

(B)     10

(C)     27

(D)     45

(E)     66

**43.** If **g(x) = log (3x+1)** then **g⁻¹(2) =**

(A)     7

(B)     10

(C)     12

(D)     26

(E)     33

**44.** If $y + 10 = (\sqrt{y} - 8)^2$ , then $\sqrt[6]{y}$ =

(A)     - 1.5

(B)     1.5

(C)     -3.375

(D)     3.375

(E)     6.225

**45.** The graph **y = 3x - 2** can be expressed as a set of parametric equations. If **x = 2 - t** and **y = f(t)** then **f(t) =**

(A)     $4 - 3t$

(B)     $3 + 4t$

(C)     $\dfrac{4 + t}{3}$

(D)     $\dfrac{3 - t}{4}$

(E)     $\dfrac{4 - t}{3}$

**46.** $\cos(x - \pi)$ =

(A)     sin x

(B)     - sin x

(C)     -cos x

(D)     - 1

(E)     1

**47.** If **f(x)** = $\dfrac{e^x}{2l\,n|x+3|}$ which value of x will make f(x) undefined ?

(A)     -9

(B)     -6

(C)     -2

(D)     0

(E)     1

SAT MATH LEVEL 1 Practice-Test

48. Find $f^{-1}(6)$

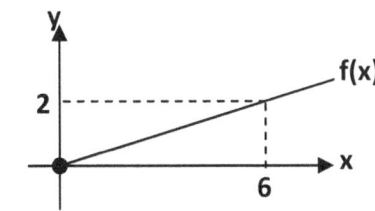

(A)    24

(B)    18

(C)    12

(D)    8

(E)    6

49.  The graph of $y = -x^2 + 12x - 9$ has ?

(A)  2 distinct real roots

(B)  2 distinct imaginary roots

(C)  2 distinct real roots and 1 imaginary root

(D)  1 imaginary root and 1 real root

(E)  1 real root only

50.  The edge of the cube has a length of 5 cm, if a rectangle is to be drawn inside the cube then what is the maximum possible area of this rectangle ?

(A)    $35\sqrt{2}$

(B)    $25\sqrt{2}$

(C)    35

(D)    25

(E)    12.5

# END OF TEST 4

**1. D**

Method: **Plug into the formula**

slope from the origin(0,0) to point A (9,12)

$$slope = \frac{12-0}{9-0} = \frac{4}{3}$$

**2. D**

Method: **Expand and Solve**

$$(x+2)^2 = (x-2)^2$$

$$x^2 + 4x + 4 = x^2 - 4x + 4$$

$$8x = 0$$

$$x = 0$$

**3. B**

Method: **Plug into the formula**

slope → m = -1

y = mx + c

y = -1x + c

x-intercepts of 5 → ( 5 , 0)

0 = -1(5) + c → c = 5

y = -1x + 5 → y + x - 5 = 0

**4. B**

Method: **Solve by drawing**

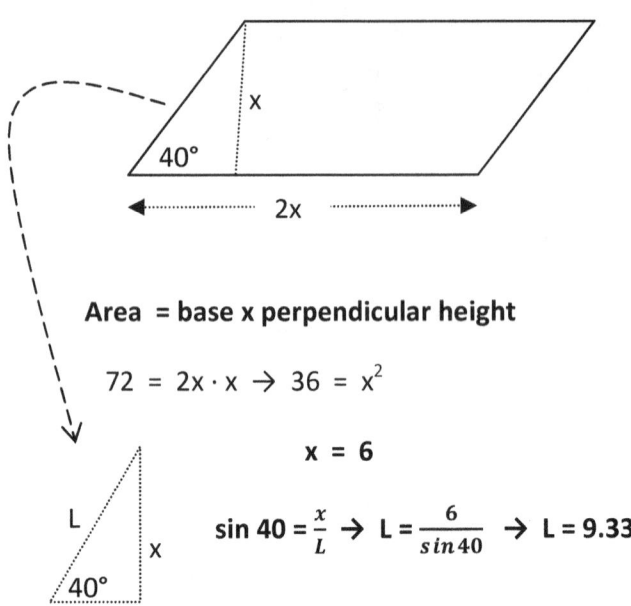

**Area = base x perpendicular height**

72 = 2x · x → 36 = $x^2$

x = 6

$\sin 40 = \frac{x}{L}$ → $L = \frac{6}{\sin 40}$ → L = 9.33

**Perimeter = 9.33 + 12 + 9.33 + 12 = 42.66**

**5. C**

Method: **Solve**

$$\frac{2\sqrt{a+b}}{3} = 1 \quad \rightarrow \quad \sqrt{a+b} = \frac{3}{2}$$

**square both side →** $\quad a + b = \frac{9}{4}$

**square again →** $\quad (a+b)^2 = \boxed{\frac{81}{16}}$

## 6. E

Method: **Graph out**

$$x^2 > |x|$$

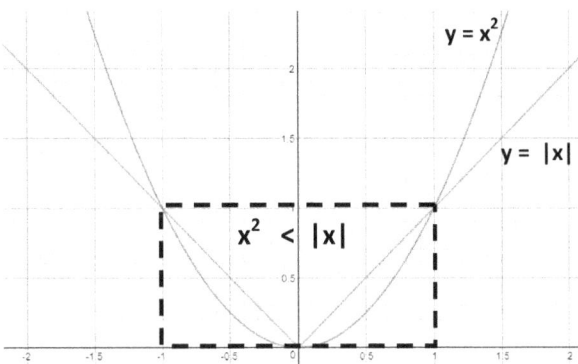

Inside the box " $x^2 < |x|$ " is true but outside the box it is the opposite.

**So we can say that " $x^2 > |x|$ " is true**

**when   x < -1 and x > 1**

## 7. A

Method: **Evaluate**

$g(x) = e^x + 1$

$$y = e^x + 1$$

swap x and y $\rightarrow$ x $= e^y + 1$

$y = ??$   $\rightarrow e^y = x - 1$

$$y = \ln(x - 1)$$

**$g^{-1}(x) = \ln(x - 1)$**

## 8. C

Method: **Use Remainder Theorem**

**$ax^2 + bx - 5$** has roots equal to **5** and **-1**

R(5) = 0 $\rightarrow$ 25a + 5b - 5 = 0   $\leftarrow$ *Equation 1*

5a + b - 1 = 0   $\leftarrow$ *Equation 1 ($\div$ 5)*

R(-1) = 0 $\rightarrow$   a - b - 5 = 0   $\leftarrow$ *Equation 2*

Solve simultaneous equation

we get  **a = 1** and  **b = -4**

therefore,   **a + b = 1 + (-4) = -3**

## 9. B

Method: **Graph and translate**

$f(x) = x^2 + 2x + 1$ $\rightarrow$ $f(x) = (x+1)^2$

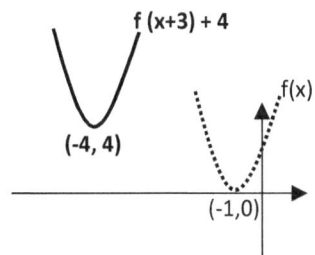

f(x) translated vertically +4 units $\rightarrow$ (-1, 4)

horizontally -3 units $\rightarrow$ **(- 4, 4)**

## 10. E

Method: **Evaluate and Simplify**

$$\lim_{x \to 0}(e^{-x} + 3)^2 = (\tfrac{1}{e^0} + 3)^2 = (1 + 3)^2 = \underline{\mathbf{16}}$$

## 11. A

Method: **Evaluate**

$1^{ST}$ , $2^{nd}$ , .......................... $5^{th}$ , $6^{th}$

Tallest , ........second to fifth..... , Shortest

1  x {           4 !           } x  1

=           **24  ways**

## 12. D

Method:  **Plug in using formula**

From the graph we know **period = 2**

and **period** $= \frac{2\pi}{b}$  →   $2 = \frac{2\pi}{b}$

$$b = \pi$$

## 13. E

Method:  **Evaluate**

Vertical asymptote means x ≠ ?

$$y = \frac{8}{x^2 - 16} \neq 0$$

so we just factor out the denominator and equate it with zero

$x^2 - 16 = 0$   → ( x - 4) (x + 4) = 0

the asymptotes are **x = 4** and **x = - 4**

## 14. B

Method: **Draw and Evaluate**

Draw out triangle

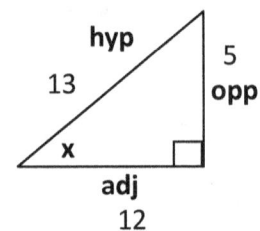

smallest angle is opposite to side length of 5

use   $\sin \theta = \frac{5}{13}$  →  **θ = 22.6°  ≈ 23**

## 15. C

Method:  **Graph out and Evaluate**

**h(t) = 5t - t$^2$**

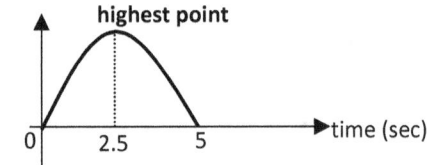

The rocket reaches highest point at **t = 2.5** s

## 16. B

Method: **Simplify**

$$(x - 3)^2 + y^2 + 7 = 16$$

$$(x - 3)^2 + y^2 = 16 - 7$$

$$(x - 3)^2 + y^2 = 9$$

We know that   $R^2 = 9$  → **R = 3**

**17. E**

Method: **Evaluate**

u and v are the domain of a function g

If $g(u) < g(v)$ then we know for sure that "u" should not be equal to "v"

so $u \neq v$

**18. C**

Method: **Draw out and Simplify**

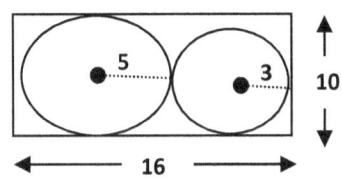

Area = 16 x 10 = **160**

**19. B**

Method: **Draw out the graph**

$y = (x - 2)^2 + 3$ → **vertex ( 2 , 3 )**

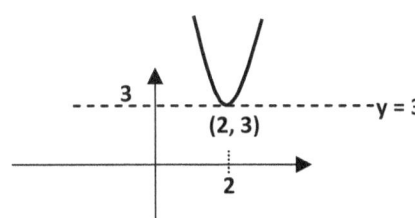

the tangent is **y = 3**

**20. C**

Method: **Solve**

$$\sqrt[5]{-\frac{32}{x^{15}}} = -16 \quad \rightarrow \quad \sqrt[5]{-\frac{2^5}{x^{15}}} = -16$$

$$\frac{-2}{x^3} = -16 \quad \rightarrow \quad x = \frac{1}{2}$$

**21. D**

Method: **Draw out and Evaluate**

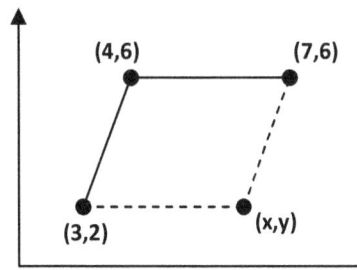

We know that slope of point (3,2) & (4,6) is same as slope of point (x,y) & (7,6)

$$\frac{6-2}{4-3} = \frac{6-y}{7-x} \quad \rightarrow \quad \frac{4}{1} = \frac{6-y}{7-x}$$

$$4 = 6 - y \quad \rightarrow \quad y = 2$$

$$1 = 7 - x \quad \rightarrow \quad x = 6$$

**So coordinate is (6,2)**

**While,**      **Area = base x height**

                 **Area = 3 x 4 = 12**

**22. D**

Method: **Evaluate using formula**

1,2,2,3,3,3,4,4,4,4.................10

The sum can be re-written as

$1 + 2^2 + 3^2 + 4^2 + \dots 10^2$

or $\sum_{n=1}^{n=10} n^2$ = **385**

**23. A**

Method: **Plug into the formula**

$3x + 2y = 12 \quad \rightarrow \quad y = -\frac{3}{2}x + 6$

Slope $= -\frac{3}{2} \quad \rightarrow$ Slope of perpendicular $= \frac{2}{3}$

The only equation with slope of 2/3 is **2x - 3y = 9**

**24. D**

Method: **Evaluate**

$2x^2 - 7x + 3 = 0$

$(2x - 1)(x - 3) = 0$

**x = 0.5** and **x = 3**

**25. E**

Method: **Graph**

range of graph **2cos(x) - 9**

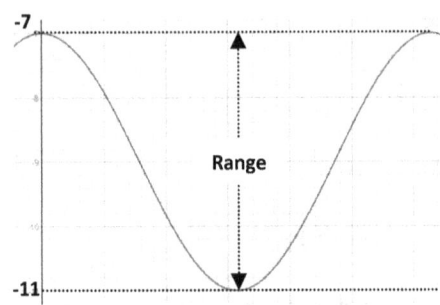

**Range : [-11, -7]**

**26. D**

Method: **Simplify**

Let the other number be ' x '

twice the other be ' 2x '

half the other be ' 0.5x ' ← smallest

Add them all $\dfrac{x + 2x + 0.5x}{3} = 50$

$2.5x = 150 \rightarrow x = 60$

the smallest → 0.5x = 0.5(60) = **30**

**27. B**

Method: **Graph out**

the graph of $y = -x^2 + 4$ and $y = -x^3$

intersect each other only **once**

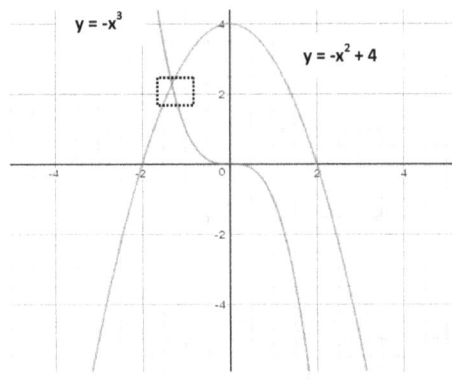

**28. D**

Method: **Plug into the formula**

$4x^2 + 2x - 1$ is divided by **(x + 1)**

using Remainder theorem plug in **x = -1** to find remainder

$R(-1) = 4(-1)^2 + 2(-1) - 1$

$R(-1) = \underline{\underline{1}}$

**29. A**

Method: **Evaluate using formula**

$U_8 = \sqrt{5} \quad \rightarrow \quad U_8 = U_1 + 7d \rightarrow \sqrt{5} = U_1 + 7d$

$\phantom{U_8 = \sqrt{5}}$ minus

$U_{10} = 3\sqrt{5} \rightarrow U_{10} = U_1 + 9d \rightarrow \underline{3\sqrt{5} = U_1 + 9d}$

$\phantom{U_{10} = 3\sqrt{5} \rightarrow} 2\sqrt{5} = 2d$

$d = \sqrt{5} \rightarrow \sqrt{5} = U_1 + 7d \rightarrow \sqrt{5} = U_1 + 7\sqrt{5}$

$U_1 = -6\sqrt{5}$

## 30. B

Method: **Simplify using formula**

We know that $\boxed{\cos^2(\theta) + \sin^2(\theta) = 1}$

$0.5\cos^2(\theta) + 0.5\sin^2(\theta) = 0.5[\cos^2(\theta) + \sin^2(\theta)]$

$= 0.5[1] = \mathbf{0.5}$

## 31. B

Method: **Evaluate using formula**

$(8 - 6i)^2 = 8^2 - 2(8)(6i) + (6i)^2$

$= \mathbf{28 - 96i}$

## 32. A

Method: **Expand using formula**

$(2 + 3x)^4 \rightarrow (3x + 2)^4 =$

$= \binom{4}{0}(3x)^4 + \boxed{\binom{4}{1}(3x)^3(2)} + \binom{4}{2}(3x)^2(2)^2 + ...$

$= ...........+ 4 (27x^3)(2) + ................$

$= ...........+ 216 x^3 + ..................$

**So the coefficient of $x^3$ is 216**

## 33. A

Method: **Expand using formula**

$(n - 2)! = (n - 3)! \cdot 4!$

$\rightarrow (n - 2)(n - 3)! = (n - 3)! \cdot 4!$

$n - 2 = 4 \cdot 3 \cdot 2 \cdot 1 \rightarrow \mathbf{n = 26}$

## 34. D

Method: **Evaluate using formula**

year $0 \rightarrow$ $ 50,000

year $1 \rightarrow$ $ (1 - 0.3) \cdot 50,000 = $ 35,000

year $2 \rightarrow$ $ (1 - 0.1) \cdot 35,000 = $ 31,500

year $3 \rightarrow$ $ (1 - 0.1) \cdot 31,500 = $ 28,350

**year $4 \rightarrow$ $ (1 - 0.1) \cdot 28,350 = $ 25,515**

## 35. C

Method: **Draw out and Evaluate**

Let $\arctan(0.25) = x$

$\tan x = 0.25 \rightarrow \mathbf{\tan x = \dfrac{1}{4}} \rightarrow \tan x = \dfrac{opp}{adj}$

Draw out triangle

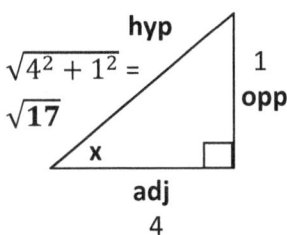

$\cos(\arctan(0.25)) = \cos(x) = \dfrac{adj}{hyp}$

$\mathbf{\cos(x) = \dfrac{4}{\sqrt{17}} = 0.97}$

**36. A**

Method: **Plug into the formula**

$\log_a 312 = 7 \rightarrow a^7 = 312$

$a = 312^{(1/7)} = $ **2.27**


**37. E**

Method: **Draw out graph**

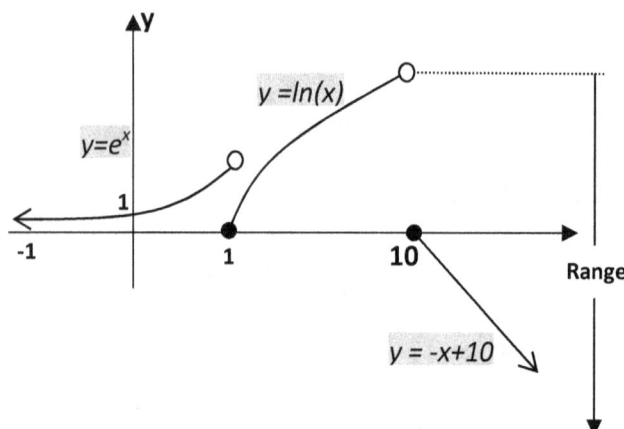

**Range : f(x) < ln(10)**


**38. E**

Method: **Evaluate**

$\{1,2,2,3,3,3,4,4,4,4\}$

Range = 4 -1 = 3 ; mean = 3 ; variance = 1.11

I. The range of the data is 4 → FALSE

II. The variance is less than the mean → TRUE

III. The mean is equal to 3 → TRUE

**So only II and III are TRUE**


**39. B**

Method: **Evaluate using ratio**

$\dfrac{Area}{Time} \rightarrow \dfrac{\pi(1.5^2)}{2\ hours} = \dfrac{3^2}{T\ hours}$

**T** = **2.5 hours**


**40. B**

Method: **Evaluate using formula**

$(x-3)^2 + (y-4)^2 = 36 \rightarrow (x-h)^2 + (y-k)^2 = R^2$

$R^2 = 36 \rightarrow$ **R = 6**

center at ( h,k ) → **(3,4)**


**41. D**

Method: **Evaluate**

$x_0 = 0.5$ and $x_{n+1} = (x_n)^2 (x_n + 2)$ then $x_4 =$

$x_1 = x_0^2 (x_0 + 2) = (0.5)^2 (0.5 + 2) = 0.625$

$x_2 = x_1^2 (x_1 + 2) = (0.625)^2 (0.625 + 2) = 1.024$

$x_3 = x_2^2 (x_2 + 2) = (1.024)^2 (1.024 + 2) = 3.171$

$x_3 \approx$ **3**

**42. D**

Method: **Evaluate**

P(red) = 1/5

P(blue) = 2/3

**P(green) = 1 - P(red) - P(blue)**

$\qquad$ = 1 - 1/5 - 2/3

$\qquad$ = **2/15**

The only number that is divisible by 15 is **45**

**43. E**

Method: **Solve by evaluation**

$g^{-1}(2)$ = ?? → means if y = 2 then x = ?

$\qquad$ g(x) = log (3x+1)

$\qquad$ 2 = $\log_{10}$ (3x+1)

$\qquad$ $10^2$ = 3x + 1

$\qquad$ 3x = 99

$\qquad$ **x = 33**

**44. B**

Method: **Evaluate by expansion**

$\sqrt[6]{y}$ = $y^{\frac{1}{6}}$ = ??

Expand $y + 10 = (\sqrt{y} - 8)^2$

y + 10 = y - $16\sqrt{y}$ + 64 → - $16\sqrt{y}$ = -54

$\sqrt{y} = \frac{27}{8}$ → $y^{\frac{1}{2}} = \frac{27}{8}$ → cube root both side

$\qquad$ $y^{\frac{1}{6}} = \frac{3}{2}$

**45. A**

Method: **Simplify**

x = 2 - t

Change parameter **x** to **t**

y = 3(**2-t**) - 2 = 6 - 3t -2

y = 4 -3t

**f(t) = 4 -3t**

**46. C**

Method: **Evaluate using formula/graph**

cos(x - π) → shift graph of cos(x) to the right by π
we will end up getting the -cos(x) graph

So, cos(x - π) = **- cos(x)**

**47. C**

Method: **Simplify**

**f(x) = $\frac{e^x}{2 \ln|x+3|}$ is undefined only when the denominator is zero.**

so, 2 ln|x+3| = 0 → ln|x+3| = 0 → x = -2

**Therefore, x = -2 will make f(x) undefined**

**48. B**

Method: **Evaluate**

We know that f(x) is linear function

with the y-intercept of 0

and slope of 1/3

So, $f(x) = \dfrac{x}{3}$ or $y = \dfrac{x}{3}$

$f^{1}(6)$ = ?? → x = ?? , y = 6

$$y = \frac{x}{3}$$

$$6 = \frac{x}{3} \rightarrow x = 18$$

therefore $f^{1}(6)$ = 18

**49. A**

Method: **Graph out**

$y = -x^2 + 12x - 9$

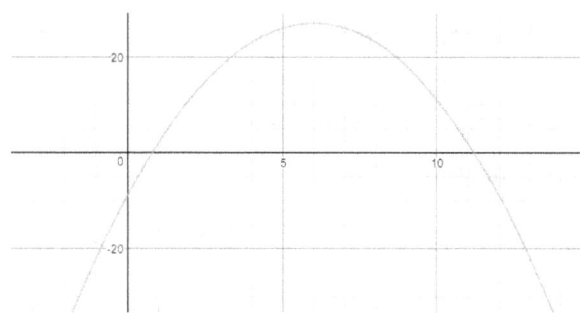

**From the graph we can conclude that it had 2 distinct real roots**

**50. B**

Method: **Evaluate using formula**

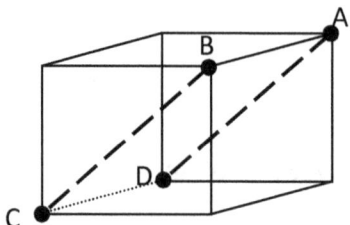

Let ABCD be the plane that contain this rectangle

we know that $BC = AD = \sqrt{5^2 + 5^2} = 5\sqrt{2}$

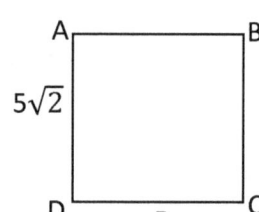

Area = $5 \times 5\sqrt{2}$

**Area = $25\sqrt{2}$**

# Score Range

| Raw Score | Conversion |
|-----------|------------|
| 45 - 50 | 800 |
| 40 - 44 | 750 - 790 |
| 36 - 39 | 720 - 740 |
| 30 - 35 | 690 - 710 |
| 25 - 29 | 650 - 680 |
| 20 - 24 | 590 - 640 |
| 16 - 19 | 540 - 580 |
| 12 - 15 | 510 - 530 |
| 9 - 11 | 450 - 500 |
| 5 - 10 | 400 - 440 |

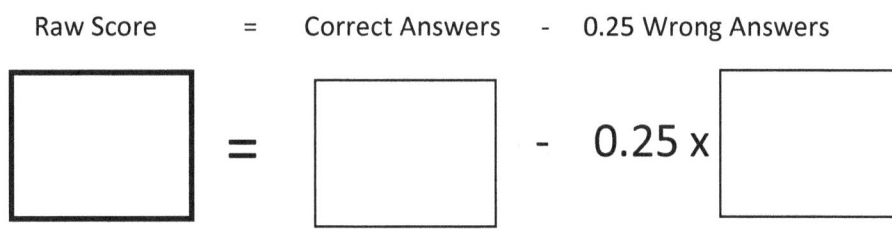

Raw Score　　=　　Correct Answers　-　0.25 Wrong Answers

$$\boxed{\phantom{XX}} = \boxed{\phantom{XX}} - 0.25 \times \boxed{\phantom{XX}}$$

1. If $\sqrt{2x} = 12$, then $x =$

(A)     296

(B)     288

(C)     144

(D)     72

(E)     64

2. $\frac{1}{x} + \frac{2}{y} + \frac{3}{z} =$

(A)     $\frac{yz + xz + xy}{xyz}$

(B)     $\frac{yz + 2xz + 3xy}{xyz}$

(C)     $\frac{xz + yz + xy}{xyz}$

(D)     $\frac{xyz}{6}$

(E)     $\frac{6}{xyz}$

3. If $f(x) = x^2 - 3x + a$, then $f(-a) - f(0) =$

(A)     $a^2 + 3a$

(B)     $a^2 + 4a$

(C)     $a^2$

(D)     $a^2 - 3a$

(E)     $a^2 - 4a$

4. The graph of $y = x^2 - 2x$ cuts the graph of $y = 8$ at

(A)     16

(B)     12

(C)     4

(D)     2

(E)     -1

5. The domain of $y = \sqrt{x + 3}$ is

(A)     $(3, \infty)$

(B)     $[-3,3)$

(C)     $[-3,\infty)$

(D)     $[-3,3) \ \cup \ (3, \infty)$

(E)     $(-\infty,-3) \cup (3, \infty)$

6. The value of $(5i - 3)^2 = a + bi$, what is the value of **b** ?

(A)     -25 - 30i

(B)     25 - 31i

(C)     16 + 31i

(D)     -16 - 30i

(E)     -16 - 31i

7. If  $\sin^2 x - \cos^2 x = 0.25$,  then  $\sin^4 x - \cos^4 x =$

(A)     0.25

(B)     0.5

(C)     1

(D)      1.25

(E)     $\sqrt{3}$

8.  What is the equation of axis of symmetry of graph $f(x) = x^2 - 6x$ ?

(A)     y = -4

(B)     y = 4

(C)     x = -3

(D)     x = 3

(E)     y = 4x - 4

9.  If a number increased by six equals to five times the square root of the  same number then what is that number ?

(A)     0

(B)     2

(C)     3

(D)     4

(E)     6

10. Which of the following represents the ratio of the sides of a right isosceles triangle ?

(A)     1:1:2

(B)     3:3:4

(C)     4:4:3

(D)     $\sqrt{2}:\sqrt{2}:1$

(E)     $\sqrt{2}:\sqrt{2}:2$

11.  The graph has an equation of

$y = a \cdot \sin(bx + c) + d$. Which of the following is the value of **b** ?

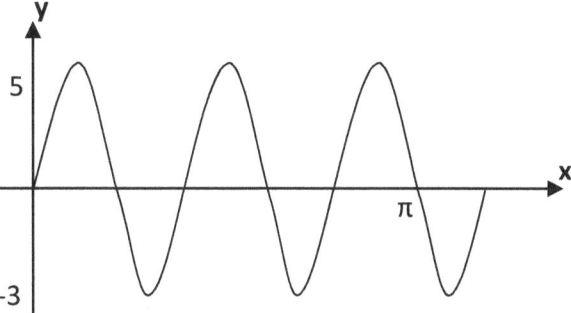

(A)     -3

(B)     -1

(C)     1

(D)     3

(E)     5

π

12. If $\log_a 4 = 2$ and $\log_b 9 = 0.5$ then  a·b  =

(A)     1

(B)     $3\sqrt{2}$

(C)     6

(D)     81

(E)     162

13.  Which value(s) of x will make the graph of rational function $y = \dfrac{5x}{x^2 - 5x}$  undefined ?

(A)     -5

(B)     0

(C)     5

(D)     -5 and 0

(E)     0  and 5

14. Line L is perpendicular bisector of line segment AB, given that A has a coordinate of (3,5) and B has a coordinate of (7,-11). Which of the following is the equation of line L?

(A)      y = x - 17

(B)      4y = 4x + 17

(C)      4y = x  - 17

(D)      y = 4x  - 17

(E)      y = x  - 4

15. If first term of arithmetic sequence is 5 and the fourth term is  35 then the sum of first ten term is equal to

(A)     410

(B)     480

(C)     500

(D)     610

(E)     720

16.  An equilateral triangle on length 5 cm is inscribed inside a square. If a square is inscribed inside a circle then what is the radius of this circle?

(A)     $\sqrt{5}$

(B)     $5\sqrt{2}$

(C)     $2\sqrt{5}$

(D)     $\dfrac{2}{\sqrt{5}}$

(E)     $\dfrac{5}{\sqrt{2}}$

17.  The diagonal of a rectangle exceeds the length by 2 cm. If the width is 10 cm, what is the area of this rectangle   ?

(A)     24

(B)     26

(C)     120

(D)     180

(E)     240

SAT MATH LEVEL 1 Practice-Test

18. If matrix A has a dimension of r by c and matrix B has a dimension of u by v then which of the following must be true ?

    I.        **AB exist if c = v**

    II.       **BA exist if v = r**

    III.     **If A + B exist then AB is possible**

(A)    I only

(B)    II only

(C)    III only

(D)    I and III only

(E)    II and III only

19. A seven sided dice is tossed two times and the results are added, what is the probability that the sum is a prime number ?

(A)    $\frac{17}{49}$

(B)    $\frac{13}{49}$

(C)    $\frac{9}{36}$

(D)    $\frac{7}{36}$

(E)    $\frac{5}{36}$

20. The formula used to calculate the amount of money (M) earned from an investment over a compound interest is **M = P(1.1)$^{0.5t}$**, where P is the initial investment and t is time. How long would it take for the investment triple its value ?

(A)    13

(B)    17

(C)    19

(D)    23

(E)    31

21. What is the distance between the origin and point $(3, -3\sqrt{2})$ ?

(A)    3

(B)    $3\sqrt{2}$

(C)    $2\sqrt{3}$

(D)    $3\sqrt{3}$

(E)    $3\sqrt{5}$

22. If f(x) = ln (g(x)) and g(x) = $e^{2x-3}$ then f(2) =

(A)    $e^1$

(B)    ln ($e^2$)

(C)    $e^{\ln 2}$

(D)    0

(E)    1

SAT MATH LEVEL 1 Practice-Test

23. A ball fallen from a shelf two meters high bounces back three-quarter of the original height. What is the total vertical distance moved by the ball ?

(A)    6  m

(B)    7  m

(C)    8  m

(D)    12 m

(E)    14 m

24. The triangle below is an isosceles triangle

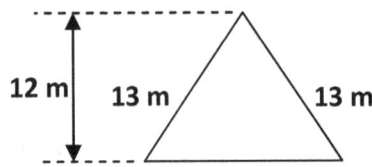

What is the perimeter of the triangle?

(A)    38

(B)    36

(C)    33

(D)    31

(E)    28

25. If $|x - 5| \leq 12$ and $|y + 2| \leq 5$ then which of the following is true of $x \cdot y$ ?

 (A)     $-15 \leq xy \leq 153$

(B)     $-51 \leq xy \leq 119$

(C)     $-60 \leq xy \leq 119$

(D)     $-119 \leq xy \leq 51$

(E)     $-119 \leq xy \leq 119$

26. If the graph of $y = (ax)^2 + 4x + 16$ has only one real root then $a =$

(A)    -4

(B)    -1

(C)    0.5

(D)    1.5

(E)    4

27. If $(2.3)^x = (3.2)^y$ then $\frac{x}{y} =$

(A)    0.716

(B)    1.396

(C)    2.312

(D)    3.221

(E)    3.698

28. The range of graph  $-3\tan(x)$  is

(A)    $[-3 , 3]$

(B)    $(-3, 0]$

(C)    $[0, 3)$

(D)    $[-\infty, 0]$

(E)    $(-\infty, \infty)$

SAT MATH LEVEL 1 Practice-Test

29. A pin code for an ATM has 4 digits, given that Patty is setting up a code for her ATM and she uses only even numbers for her pin except zero. How many pin code combinations are there ?

(A)     16

(B)     64

(C)     256

(D)     625

(E)     1000

30. If the equilateral triangle has an area of $\frac{\sqrt{3}}{4}$ then what is its perimeter ?

(A)     3

(B)     4

(C)     5

(D)     $\frac{3\sqrt{3}}{4}$

(E)     $\frac{5\sqrt{3}}{4}$

31. The kinetic energy varies directly as the square of the speed, if a car was moving with the speed of 40 km/hr and it posses kinetic energy of "K" then in terms of "K" what will its kinetic energy be if its speed was 120 km/hr ?

(A)     9K

(B)     4K

(C)     3K

(D)     0.25K

(E)     0.133K

32. Given that

$$f(x) = \begin{cases} 4 & , x \le 0 \\ -x^2 + 4 & , 0 < x < 2 \\ x - 2 & , x > 2 \end{cases}$$

What is the range of f(x) ?

(A)     $0 < f(x) < \infty$

(B)     $-\infty < f(x) \le 0$

(C)     $-\infty < f(x) < 0$ and $0 < f(x) < \infty$

(D)     $-\infty < f(x) < 2$ and $2 \le f(x) < \infty$

(E)     All real numbers

33. The value of $3i^2 - i^4 + 4i^6$ is equal to

(A)     - 8

(B)     - 6

(C)     0

(D)     6

(E)     8

34. Which of the following statement must be true about the data set **{3,5,x,7,9}** ? Given that the mean is 7.

I. The range of the data is 6

II. The mode is 7

III. The median is 7

(A)    I only

(B)    II only

(C)    III only

(D)    I and II only

(E)    II and III only

35. Which of the following is the equation of the circle with radius 2 and center at (3, 4) ?

(A)    $(x - 3)^2 - (y - 4)^2 = 4$

(B)    $(x + 3)^2 - (y + 4)^2 = 4$

(C)    $(x + 3)^2 + (y + 4)^2 = 2$

(D)    $(x - 3)^2 + (y - 4)^2 = 2$

(E)    $(x - 3)^2 + (y - 4)^2 = 4$

36. What is the unit digit of $7^{777}$ ?

(A)    1

(B)    3

(C)    7

(D)    9

(E)    0

37. If **f(x) = ln (cos(x))** , $0 < x < 2\pi$,

then **f(0) =**

(A)      - 1

(B)      0

(C)      1

(D)      e

(E)      undefined

38. If **a$^{-3}$ = -216** then **a =**

(A)    $\frac{-1}{6}$

(B)    $\frac{1}{6}$

(C)    -6

(D)    6

(E)    36

39. The sum of first 30 multiple of 11 is how much greater than the sum of first 30 multiples of 5 ?

(A)    2325

(B)    2790

(C)    3225

(D)    4615

(E)    5115

SAT MATH LEVEL 1 Practice-Test

40. The exterior angle of a regular hexagon equal to

(A)    20

(B)    30

(C)    60

(D)    100

(E)    120

41. If $(\log_2 x)(\log_2 2) = 4$ then $x =$

(A)    2

(B)    4

(C)    16

(D)    36

(E)    81

42. The operation $x\,\$\,y$ is defined by taking all the prime numbers between x and y then multiply them to obtain the result. What is the result of $12\,\$\,23$ ?

(A)    221

(B)    323

(C)    2431

(D)    4199

(E)    4641

43. If $\csc\theta = \sqrt{2}$ then $\tan\theta =$

(A)    0.5

(B)    $\frac{2}{3}$

(C)    1

(D)    $\sqrt{3}$

(E)    $\frac{\sqrt{2}}{\sqrt{3}}$

44. If $a^2 \cdot b^2 < 0$ then which of the following is true ?

(A)    a is real negative number

(B)    b is real negative number

(C)    a and b are real negative number

(D)    a is an imaginary number

(E)    a and b are imaginary numbers

45. A cone has a diameter of 12 cm and it holds 905 cm$^3$ liquid.  To the nearest centimeter, what is the height of this cone?

(A)    2

(B)    9

(C)    16

(D)    24

(E)    75

46. 27. Which of the following is the solution set for the equation $12 - x \le 3x + 4$

(A)

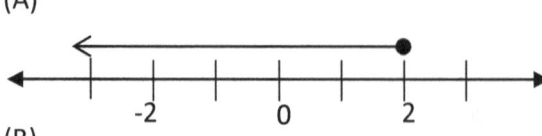

(B)

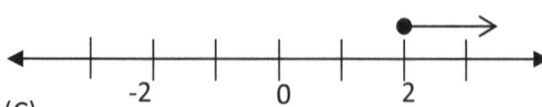

(C)

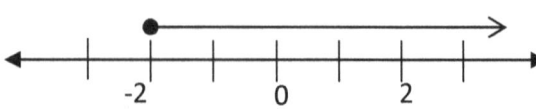

(D)

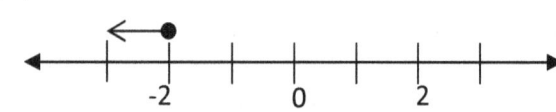

(E)

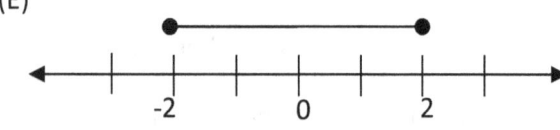

47. The graph of $f(x) = |x|$ is translate up 5 units, reflected once on x-axis and translate to the right 4 units. Which of the following function represents the translations of $f(x)$ ?

(A)  $|x - 4| + 5$

(B)  $|x + 4| - 5$

(C)  $|x - 4| + 5$

(D)  $-|x - 4| - 5$

(E)  $-|x + 4| - 5$

48. If $(0.5x)! = (x-2)!$ , then $x =$

(A)  10

(B)  9

(C)  8

(D)  6

(E)  4

49. If $b^{0.5a} = 16$, 'a' and 'b' are positive integers greater than 1. What a possible value of 'a' ?

(A)  2

(B)  4

(C)  6

(D)  8

(E)  10

50.  A group of students decided to divide the $900 cost of a trip equally among themselves. When three of the students decided not to go on the trip, those remaining still divided the $900 cost equally, but each student's share of the cost increased by $50. How many students were in the group originally?

(A)  9

(B)  7

(C)  6

(D)  5

(E)  4

# END OF TEST 5

SAT MATH LEVEL 1 Practice-Test

**1. D**

Method: **Solve**

$\sqrt{2x} = 12$ → square both side

  2x = 144

    **x = 72**

**2. B**

Method: **Simplify**

$$\frac{1}{x} + \frac{2}{y} + \frac{3}{z} = \frac{yz + 2xz + 3xy}{xyz}$$

**3. A**

Method: **Evaluate the function**

**f(x) = x² - 3x + a,**

f(0) = a

f(-a) = (-a)² - 3(-a) + a = a² + 4a

**f(-a) - f(0)** = a² + 4a - a

**f(-a) - f(0)** = **a² + 3a**

**4. D**

Method: **Simplify and Evaluate**

y = x² - 2x   and   y = 8  cut each other

equate them    x² - 2x = 8

             x² - 2x - 8 = 0

           (x - 4)( x + 2) = 0

       **x = 4**   or   x = -2

**5. C**

Method: **Evaluate**

**y = $\sqrt{x + 3}$**

We know that a square root graph must have the input (x+3) value of greater than or equal to zero. Therefore,

          **x + 3 ≥ 0**

           **x ≥ -3**

**Domain : [-3,∞)**

**6. D**

Method: **Expand and simplify**

*(5i - 3)² = (5i)² - 2(5i)(3) + (3²)*

     *= -25 - 30i + 9*

     *= **-16 - 30i***

**7. A**

Method: **Factor and Evaluate**

$\sin^4 x - \cos^4 x = (\sin^2 x - \cos^2 x)(\sin^2 x + \cos^2 x)$

$\qquad = \qquad (0.25) \qquad (1)$

$\qquad = \qquad \mathbf{0.25}$

**8. D**

Method: **Complete the square and Graph**

$f(x) = x^2 - 6x$ → we use vertex equation

vertex $\mathbf{x} = \dfrac{-b}{2a}$ ( this is the axis of symmetry eq.)

$\qquad x = \dfrac{-(-6)}{2(1)} \rightarrow \qquad \mathbf{x = 3}$

**9. D**

Method: **Evaluate**

a number increased by six → **x + 6**

five times the square root of a number → $5\sqrt{x}$

$\qquad x + 6 = 5\sqrt{x}$

$\qquad x - 5\sqrt{x} + 6 = 0$

$\qquad (\sqrt{x} - 3)(\sqrt{x} - 2) = 0$

$\qquad \sqrt{x} = 3 \rightarrow x = 9 \qquad$ or

$\qquad \sqrt{x} = 2 \rightarrow \boxed{x = 4}$

**10. E**

Method: **Simplify the drawing**

**Draw out the right isosceles triangle**

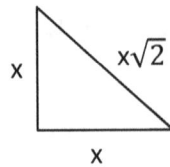

$\qquad$ ratio $\qquad \rightarrow x : x : x\sqrt{2}$
$\qquad$ simplify $\rightarrow 1 : 1 : \sqrt{2}$

multiply by $\sqrt{2} \rightarrow \sqrt{2} : \sqrt{2} : \mathbf{2}$

**11. E**

Method: **Solve by translation**

**b** represents **the period of the graph**

We know that the period of the graph is $\dfrac{2\pi}{5}$

formula $\rightarrow$ Period $= \dfrac{2\pi}{b}$

$\qquad \dfrac{\pi}{5} = \dfrac{2\pi}{b}$

$\qquad \boxed{b = 5}$

**12. E**

Method: **Evaluate**

$\log_a 4 = 2 \quad \rightarrow a^2 = 4 \rightarrow a = 2$

$\log_b 9 = 0.5 \quad \rightarrow b^{0.5} = 9 \rightarrow b = 81$

$\quad a \cdot b = 2(81) = \underline{162}$

## 13. E

Method: **Evaluate**

$$y = \frac{5x}{x^2 - 5x} \rightarrow y = \frac{5x}{x(x-5)}$$

**denominator cannot be equal to zero**

therefore,    $x \cdot (x-5) \neq 0$

$$x \neq 0 \text{ and } x \neq 5$$

## 14. C

Method: **Evaluate using formula**

Find midpoint $\rightarrow (\frac{7+3}{2}, \frac{-11+5}{2})$ = **(5, -3)**

$Slope_{AB} = \frac{-11-5}{7-3} = \frac{-16}{4} = -4$

$\mathbf{m = Slope_L} = \frac{-1}{Slope_{AB}} = \frac{1}{4}$

$y = mx + c$

Put **(5, -3)** and $\mathbf{m = \frac{1}{4}}$

$-3 = \frac{1}{4}(5) + c$

$c = \frac{-17}{4}$

$y = \frac{1}{4}x - \frac{17}{4}$

**4y = x - 17**

## 15. C

Method: **Plug into the formula**

$U_1 = 5$ , $U_4 = 35$

$U_4 = U_1 + 3d \rightarrow 35 = 5 + 3d \rightarrow d = 10$

$U_{10} = U_1 + (10 - 1)d = 5 + 9(10) = 95$

$S_{10} = \frac{10}{2}(U_1 + U_{10}) = 5(5 + 95) = \mathbf{500}$

## 16. E

Method: **Draw out and Evaluate**

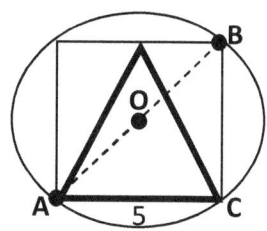

Using Pythagoras $\rightarrow$

$$(2R)^2 = 5^2 + 5^2$$

$$R^2 = \frac{50}{4}$$

$$R = \frac{5}{\sqrt{2}}$$

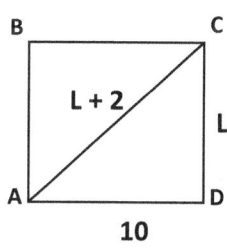

## 17. E

Method: **Draw out and evaluate**

$(L+2)^2 = L^2 + 10^2 \rightarrow L = 24$

Area = Base x Length

**Area = 10 x 24 = 240**

## 18. B

Method: **Plug-in choices**

> I.   AB exist if c = v → FALSE
>
> C should equal to u

II.   BA exist if v = r → TRUE

> III.   If A + B exist then AB is possible → FALSE
>
> If A has a dimension of 2 x 3 and B is a dimension of 2 x 3 as well then AB is not possible

## 19. A

Method: **Evaluate**

Sum of prime numbers are

2, 3, 5, 7, 11 and 13 only

| Numbers | Sum of |
|---------|--------|
| 2 | (1,1) |
| 3 | (1,2) , (2,1) |
| 5 | (1,4), (2,3), (3,2), (4,1) |
| 7 | (1,6) , (2,5), (3,4), (4,3),(5,2),(6,1) |
| 11 | (4,7), (7,4) |
| 13 | (6,7), (7,6) |

Total possibilities of prime sum = 17

P(sum of square number) = $\frac{17}{49}$

## 20. D

Method: **Solve**

triple means  M = 3P

$$M = P(1.1)^{0.5t}$$

$$3P = P(1.1)^{0.5t}$$

$$3 = (1.1)^{0.5t}$$

$$\ln(3) = \ln(1.1)^{0.5t}$$

$$0.5t = \frac{\ln(3)}{\ln(1.1)}$$

$$t = \textbf{23 years}$$

## 21. D

Method: **Plug into the formula**

distance from the origin to point $(3 , -3\sqrt{2})$

distance = $\sqrt{(3-0)^2 + (-3\sqrt{2}-0)^2}$

distance = $\sqrt{27}$ = $\mathbf{3\sqrt{3}}$

## 22. E

Method: **Plug into the formula**

$f(x) = \ln(g(x))$  and  $g(x) = e^{2x-3}$

$f(2) = \ln(g(2))$ → $g(2) = e^{2(2)-3} = e$

$f(2) = \ln( e )$

$\mathbf{f(2) = \underline{1}}$

**23. E**

Method: **Plug into the formula**

| | | | |
|---|---|---|---|
| **First drop** | = | **2 m** | |
| **Second drop** | = | **2(3/4)** | = **1.5 m** |
| **Third drop** | = | **1.5(3/4)** | = **1.125 m** |

⋮

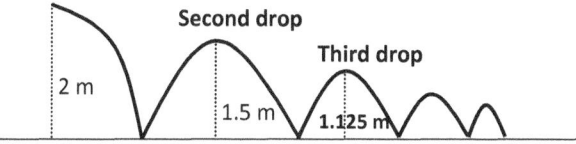

First drop

Second drop

Third drop

2 m

1.5 m

1.125 m

These are geometric sequences (with r = 3/4)

Falling sequence = 2, 1.5, 1.125, ....

Bouncing sequence = 1.5, 1.125, ....

Total distance = Sum of falling + Sum of Bouncing
to infinity term       to infinity term

$$= \frac{2}{1-3/4} + \frac{1.5}{1-3/4} = 8 + 6 = \textbf{14 m}$$

**24. B**

Method: **Simplify**

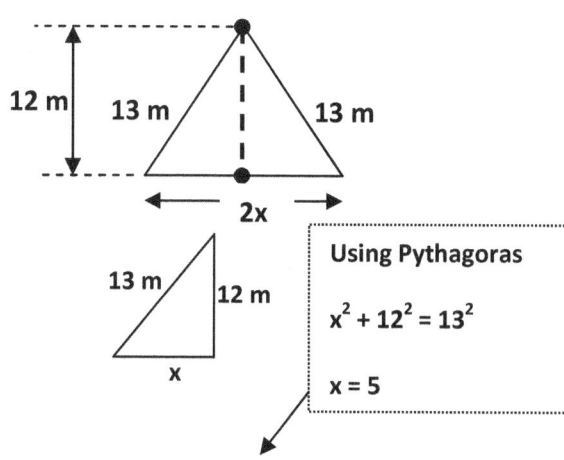

12 m   13 m   13 m

2x

13 m   12 m

x

**Using Pythagoras**

$x^2 + 12^2 = 13^2$

x = 5

Perimeter = 13 + 13 + 2x = **36**

**25. D**

Method: **Solve**

$|x - 5| \leq 12$

case 1: $x - 5 \leq 12 \rightarrow x \leq 17$

case 2: $x - 5 \geq -12 \rightarrow x \geq -7$

$-7 \leq x \leq 17$

$|y+2| \leq 5$

case 1: $y+2 \leq 5 \rightarrow y \leq 3$

case 2: $y+2 \geq -5 \rightarrow y \geq -7$

$-7 \leq y \leq 3$

Therefore,

$-7(17) \leq xy \leq 3(17)$

$\textbf{-119} \leq \textbf{xy} \leq \textbf{51}$

**26. C**

Method: **Evaluate**

We know from quadratic equation that if the graph has one root its Discriminant should equal to **zero**.

$$\text{Discriminant} = (4^2) - 4(a^2)(16)$$

$$0 = 16 - 64a^2$$

$$a^2 = \frac{1}{4} \rightarrow a = \frac{1}{2}$$

**27. B**

Method: **Evaluate**

$$(2.3)^x = (3.2)^y$$

$$\ln(2.3)^x = \ln(3.2)^y$$

$$x \cdot \ln(2.3) = y \cdot \ln(3.2)$$

$$\frac{x}{y} = \frac{\ln(3.2)}{\ln(2.3)}$$

$$\frac{x}{y} = 1.39$$

**28. E**

Method: **Draw out graph**

$$y = -3\tan(x)$$

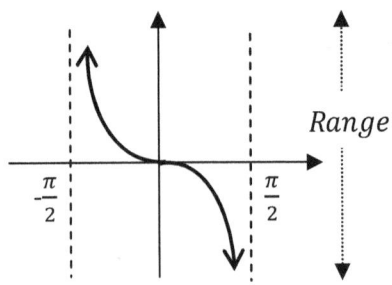

**Range: $(-\infty, \infty)$**

**29. C**

Method: **Create box of possibilities**

digits possible are {2, 4, 6, 8} only 4 possibilities

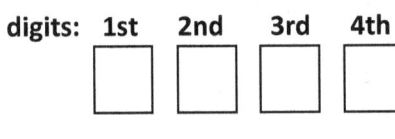

$$4 \times 4 \times 4 \times 4$$

**total possibilities = $4^4$ = 256**

**30. A**

Method: **Draw out and Evaluate**

**Area of triangle $= \frac{1}{2} x \cdot x \cdot \sin 60°$**

$$\frac{\sqrt{3}}{4} = \frac{1}{2} \cdot x^2 \cdot \frac{\sqrt{3}}{2}$$

$$x = 1$$

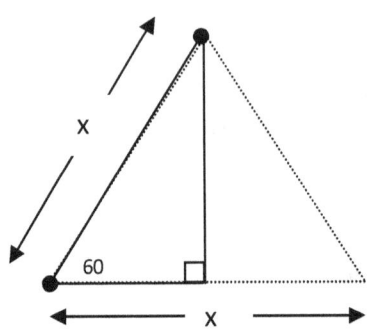

**Perimeter = 3x = 3(1) = <u>3</u>**

**31. A**

Method: **Evaluate**

Let A be a constant that link energy and speed

energy is directly proportional to speed→ **energy = A·Speed$^2$**

$$\text{energy} = A \cdot \text{Speed}^2 \rightarrow A = \frac{energy}{speed^2}$$

| Energy | Speed$^2$ | A (constant) |
|--------|-----------|--------------|
| K | $40^2$ | $\frac{K}{40^2}$ |
| ?? | $120^2$ | $\frac{K}{40^2}$ |

energy = A·Speed$^2$

$$?? = \frac{K}{40^2} \cdot 120^2$$

**?? = <u>9K</u>**

**32. A**

Method: **Graph**

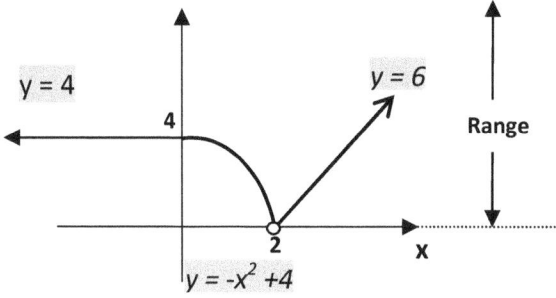

**Range: 0 < f(x) < ∞**

**33. A**

Method: **Simplify and Plug into the formula**

| $3i^2 - i^4 + 4i^6$ | = | $3(-1) - (1) + 4(-1)$ |
|---|---|---|
| | = | $-3 - 1 - 4$ |
| | = | **- 8** |

**34. C**

Method: **Solve**

$\frac{(3+5+x+7+9)}{5} = 7$ → x = 35 - 24 → **x = 11**

I. The range of the data is 6 → FALSE

Range = 11 - 3 = 8

II. The mode is 7 → FALSE

III. The median is 7 → TRUE

Rearrange the numbers {3,5,7,9,11}

median = 7

**35. E**

Method: **Plug into the formula**

equation of circle → $(x - h)^2 + (y - k)^2 = r^2$

(h , k) → (3,4)

Radius → r = 2

$(x - 3)^2 + (y - 4)^2 = 2^2$

$(x - 3)^2 + (y - 4)^2 = 4$

**36. C**

Method: **Evaluate**

$7^1, 7^2, 7^3, 7^4, 7^5, 7^6, 7^7, 7^8, .....$

= 7, 49, 343, 2401, 16807, 117649, ...

units digits → 7 , 9 , 3 , 1, 7 , 9 , 3 ,1 .... The pattern repeats every 4 times

$7^{777}$ → power 777 ÷ 4 = $194\frac{1}{4}$ → remainder of 1

**Remainder of 1 means the position will be on the first term which is 7.**

**37. C**

Method: **Evaluate**

| f(x) | = | ln(cos(x)) |
|---|---|---|
| f(0) | = | ln(cos(0)) |
| | = | ln ( 1 ) |
| | = | **0** |

**38. A**

Method: **Solve**

$$a^{-3} = -216 \rightarrow \frac{1}{a^3} = -216 \rightarrow \frac{1}{-216} = a^3$$

$$a = \sqrt[3]{\frac{1}{-216}} \rightarrow a = \frac{-1}{6}$$

**39. B**

Method: **Solve**

> sum of first 30 multiple of 11
>
> 11, 22, ........... ,(11x30) = 330
>
> $\sum_{i=1}^{30} 11i = \frac{30}{2}(11 + 330) = 5115$

> sum of first 30 multiples of 5
>
> 5, 10 , .......... ,(5 x 30) = 150
>
> $\sum_{i=1}^{30} 5i = \frac{30}{2}(5 + 150) = 2325$

**Differences = 5115 - 2325 = <u>2790</u>**

**40. A**

Method: **Evaluate using formula**

Sum of exterior angle of a polygon is 360

Hexagon has 6 sides $\rightarrow$ n = 6

Each exterior angle $= \frac{360}{n}$

$$= \frac{360}{6} = \underline{60°}$$

**41. C**

Method: **Evaluate using formula**

$$(\log_2 x)(\log_2 2) = 4$$

$$(\log_2 x)\ (1) = 4$$

$$2^4 = x$$

$$x = 16$$

**42. D**

Method: **Evaluate**

12 $ 23 = 13 x 17 x 19 = **4199**

**43. C**

Method: **Draw out triangle and Simplify**

$$\csc\theta = \frac{\sqrt{2}}{1} = \frac{hyp}{opp}$$

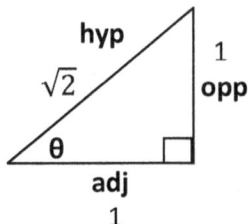

$$\tan\theta = \frac{opp}{adj} = 1$$

**44. D**

Method:  **Solve by evaluating**

$$a^2 \cdot b^2 < 0$$

one of the square must be negative numbers

Case 1: Let $a^2$ be a negative number then $b^2$ must be a positive number

$a^2 = -1$  and  $b^2 = 1 \rightarrow$  $a^2 \cdot b^2 < 0$

$$(-1)\,(1) < 0$$

$$-1 \ < 0 \ \textbf{TRUE}$$

Case 2: Let  $b^2$ be a negative number then $a^2$ must be a positive number

$a^2 = 1$  and  $b^2 = -1 \rightarrow$  $a^2 \cdot b^2 < 0$

$$(1)\,(-1) < 0$$

$$-1 \ < 0 \ \textbf{TRUE}$$

**Therefore, one of the square must be a negative number.**

$$a^2 = -1 \rightarrow a = \sqrt{-1} = \ i$$

**Both of the squares cannot be a negative number only one of them can be. So one of them is an imaginary number.**

**45. D**

Method:  **Evaluate by using formula**

diameter $= 12 \rightarrow$  radius $= 6$

| **Volume of cone** | = | $\dfrac{\pi r^2 h}{3}$ |
|---|---|---|
| 905 | = | $\dfrac{\pi 6^2 h}{3}$ |
| h | = | **24** |

**46. B**

Method:  **Solve**

$$12 - x \leq 3x + 4$$

$$-4x \leq -8$$

$$\mathbf{x \geq 2}$$

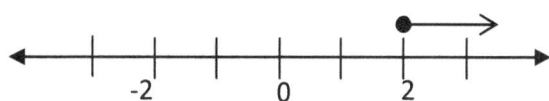

**47. D**

Method:  **Translate with graph algebraically**

$f(x) = |x| \ \rightarrow$  translate up 5 units  $\rightarrow \ |x| + 5$

$|x| + 5 \ \rightarrow$ reflected on x-axis   $\rightarrow -|x| - 5$

$-|x| - 5 \ \rightarrow$ translate 4 units right $\rightarrow \mathbf{-|x - 4| - 5}$

## 48. E

Method: **Plug in the choice**

| | x | 0.5x | x - 2 | Equal |
|---|---|---|---|---|
| A | 10 | 5 | 8 | NO |
| B | 9 | 4.5 | 7 | NO |
| C | 8 | 4 | 6 | NO |
| D | 6 | 3 | 4 | NO |
| E | 4 | 2 | 2 | YES |

## 49. D

Method: **Evaluate**

Since $a > 1$ and $b > 1$

$$b^{0.5a} = 16$$

$$b^{0.5a} = 2^4$$

equating the terms:

$$b = 2 \quad \text{and} \quad 0.5a = 4$$

$$a = \underline{8}$$

## 50. A

Method: **Evaluate algebraically**

At first each pays $\rightarrow \dfrac{900}{x}$

Let 'x' be the total number of students

Now three are not going each would have to pay more $\rightarrow \dfrac{900}{x-3}$

Each now pays 50 more , equate the equations:

$$\frac{900}{x} + 50 = \frac{900}{x-3}$$

$$x^2 - 3x + 54 = 0$$

$$(x - 9)(x + 6) = 0$$

$$\mathbf{x = 9} \quad \text{or} \quad x = -6$$

**So there were originally 9 people in the group**

# Score Range

| Raw Score | Conversion |
|---|---|
| 47 - 50 | 800 |
| 42 - 46 | 750 - 790 |
| 38 - 41 | 720 - 740 |
| 34 - 37 | 690 - 710 |
| 30 - 33 | 650 - 680 |
| 26 - 29 | 590 - 640 |
| 23 - 27 | 540 - 580 |
| 18 - 22 | 510 - 530 |
| 13 - 17 | 450 - 500 |
| 7 - 11 | 400 - 440 |

Raw Score     =     Correct Answers     -     0.25 Wrong Answers

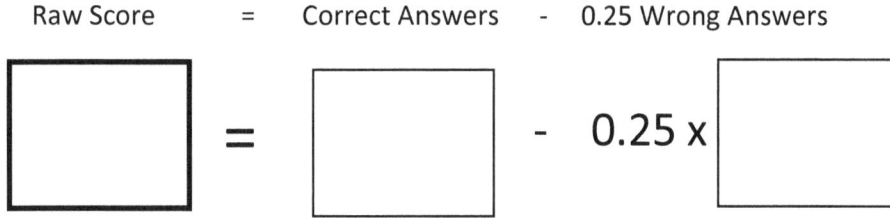

=      - 0.25 x

# Formula:

## Arithmetic

$Mean = \dfrac{X1+X2+X3+..+Xn}{n}$ , Median = middle number value , Mode = occur most

$Variance = \dfrac{\sum(X-\mu)^2}{N}$ , Standard deviation $= \sqrt{Variance}$

$Percent\ Change = \dfrac{(New-Original)\times100\%}{Original}$

$Simple\ Interest = \dfrac{(Principal\ x\ Rate\ x\ Time)}{100}$

Compound Interest: Total money earned $= principal\left(1+\dfrac{rate}{100}\right)^{time}$

Probability: Probability of an event $= \dfrac{number\ of\ favorable\ outcomes}{number\ of\ possible\ outcomes}$

## Geometry (2D)

$Area\ of\ triangle = \dfrac{1}{2} \times Base \times Height$

$Area\ of\ square = Side^2$

$Area\ of\ rectangle = Length\ x\ Width$

$Area\ of\ parallelogram = Base\ x\ Height$

$Area\ of\ trapezium = \dfrac{1}{2} \times (Base1 + Base2) \times Height$

$Sum\ of\ Interior\ Angle = (n-2)x180$

where "n" the number of sides or angles

## Geometry (3D)

Volume of Cube = edge$^3$

Volume of Rectangular  solid = length x width x height

Volume of Square  pyramid = $\frac{1}{3}$ x ( base edge)$^2$ x height

Volume of Cylinder = $\pi$ x radius$^2$ x height

Volume of Cone = $\frac{1}{3}$x $\pi$ x radius$^2$ x height

## Algebra

*Linear Equation:*

$y = mx + c$

"*c*" is a *y*-intercept, where the graph crosses y-axis

"*m*" is a gradient or slope

$$\text{Slope} = \frac{\Delta Y}{\Delta X} = \frac{y_2 - y_1}{x_2 - x_1}$$

*Quadratic Equation:*

$(x + y)^2 = x^2 + 2xy + y^2$

$(x - y)^2 = x^2 - 2xy + y^2$

$(x + y)(x - y) = x^2 - y^2$

$$ax^2 + bx + c = 0$$

*Quadratic formula:*
$$x = \frac{-b \pm \sqrt{b^2 - 4ac}}{2a}$$

**Equation of parabola:**

| | x-axis | y-axis |
|---|---|---|
| | $y = a(x - h)^2 + k$ | $x = a(y - k)^2 + h$ |
| Axis of Symmetry: | $x = h$ | $y = k$ |
| Vertex: | $(h, k)$ | $(h, k)$ |
| Focus: | $\left(h, k + \frac{1}{4a}\right)$ | $\left(h + \frac{1}{4a}, k\right)$ |
| Directrix: | $y = k - \frac{1}{4a}$ | $x = h - \frac{1}{4a}$ |

**Equation of circle:**

$$(x - h)^2 + (y - k)^2 = R^2$$

Center of the circle at **(h, k)**

Radius equals to **R**

**Equation of ellipse:**

| | x-axis | y-axis | |
|---|---|---|---|
| | $\dfrac{(x-h)^2}{a^2} + \dfrac{(y-k)^2}{b^2} = 1$ | $\dfrac{(x-h)^2}{b^2} + \dfrac{(y-k)^2}{a^2} = 1$ | *For, a > b* |
| Center: | $(h, k)$ | $(h, k)$ | |
| Semi-Major axis: | $a$ | $a$ | |
| Semi-Minor axis: | $b$ | $b$ | |

**Equation of hyperbola:**

| | x-axis | y-axis | |
|---|---|---|---|
| | $\dfrac{(x-h)^2}{a^2} - \dfrac{(y-k)^2}{b^2} = 1$ | $-\dfrac{(x-h)^2}{b^2} + \dfrac{(y-k)^2}{a^2} = 1$ | *For, $c^2 = a^2 + b^2$* |
| Center: | $(h, k)$ | $(h, k)$ | |
| Focus : | $(h \pm c, k)$ | $(h, k \pm c)$ | |
| Vertex : | $(h \pm a, k)$ | $(h, k \pm a)$ | |
| Asymptotes: | $y - k = \pm \dfrac{b}{a}(x - h)$ | $y - k = \pm \dfrac{a}{b}(x - h)$ | |